Contents

CW01500284

Science in the Third Reich

German Historical Perspective Series
General Editors:
Gerhard A. Ritter and Anthony J. Nicholls

German Historical Perspectives/XII

Science in the Third Reich

Edited by
MARGIT SZÖLLÖSI-JANZE

Oxford • New York

First published in 2001 by
Berg
Editorial offices:
150 Cowley Road, Oxford, OX4 1JJ, UK
838 Broadway, Third Floor, New York, NY 10003-4812, USA

Berg is the imprint of Oxford International Publishers Ltd.

Library of Congress Cataloging-in-Publication Data

A catalogue record for this book is available from the Library of Congress.

British Library Cataloguing-in-Publication Data

A catalogue record for this book is available from the British Library

ISBN 1 85973 416 2 (Cloth)
       1 85973 421 9 (Paper)

Typeset by JS Typesetting, Wellingborough, Northants.
Printed in the United Kingdom by Biddles Ltd, Guildford and King's Lynn.

Editorial Preface

The purpose of this series of books is to present the results of research by German historians and social scientists to readers in English-speaking countries. Each of the volumes has a particular theme that will be handled from different points of view by specialists. The series is not limited to the problems of Germany but will also involve publications dealing with the history of other countries, with the general problems of political, economic, social and intellectual history as well as international relations and studies in comparative history.

We hope the series will help to overcome the language barrier that experience has shown obstructs the rapid appreciation of German research in English-speaking countries.

The publication of the series is closely associated with the German Visiting Fellowship at St Antony's College, Oxford, which has existed since 1965, having been originally funded by the Volkswagen Stiftung, later by the British Leverhulme Trust, by the Ministry of Education and Science in the Federal Republic of Germany, and starting in 1990, by the Stifterverband für die Deutsche Wissenschaft with special funding from C. & A. Mode Düsseldorf. Each volume is based on a series of seminars held in Oxford, which has been conceived and directed by the Visiting Fellow and organized in collaboration with St Antony's College.

The editors wish to thank the Stifterverband für die Deutsche Wissenschaft for meeting the expenses of the original lecture series and for generous assistance with the publication. They hope that this enterprise will help to overcome national introspection and to further international academic discourse and co-operation.

Gerhard A. Ritter   Anthony J. Nicholls

MARGIT SZÖLLÖSI-JANZE

National Socialism and the Sciences: Reflections, Conclusions and Historical Perspectives

I

In 1935 the Reich Leader of the SS, Heinrich Himmler, founded a scientific research institute, to be known as Das Ahnenerbe (literally: 'Ancestral Heritage'). He defined its task as the 'implementation of fundamental and most valuable research, which nevertheless went wholly or partially unrecognized, or was even persecuted by the official sciences' (Himmler to Göring, 19 March 1937, cit. in Ackermann 1970: 45). Among those studies rejected by institutionalized science and promoted by the SS was the so-called *Welteislehre* or *Glazial-Kosmogonie* ('World Ice Cosmogony') (see Nagel 1991: 11–27; Nagel 1994). This theory traced the formation of the cosmos, its suns and stars back to the existence of never-ending ice throughout the universe, which, in its struggle with blazing giant suns, formed the driving power of all cosmic changes. Our solar system, with all its planets, was said to be the result of the collision of a giant sun with a smaller celestial body of ice. The surface of the moon and all the planets, with the exception of the Earth, were encased in a shell of ice. Cosmic ice formed the Milky Way, shooting stars, hailstones, or, when these evaporated, rainclouds. According to the theory, ice was the matter from which the world was formed.

The *Welteislehre* claimed to provide comprehensive, scientific explanations for all observable phenomena, from the beginning to the end of the world. From the eternal struggle of bodies of heat

1

against bodies of ice, for example, it derived a law which suggested that through the restraining resistance of the supposed ether in the cosmos, all smaller celestial bodies would be 'caught' as the moons of the next largest planet and, entering ever tighter spiralling orbits, would eventually collide with it. The sun had already 'caught' ten planetoids, the Earth three. As evidence for volcanic eruptions, earthquakes and floods, which this caused on the Earth, the *Welteislehre* cited the Bible, in particular the Flood in Genesis and the visions of the Apocalypse. The break down of moons (*Mondniederbrüche*) explained the extinction of the dinosaurs, the ice-ages, the emergence of coal-seams and the sinking of the legendary land of Atlantis into the ocean. According to the *Welteislehre* the Earth would eventually, in the distant future, 'capture' Mars and together they would plunge into the sun.

Is this 'Nazi science'? Some things appear to suggest that that is the case. Not only Himmler, but also Hitler, Baldur von Schirach and perhaps Hermann Göring are supposed to have been convinced disciples of glacial cosmogony (Ackermann 1970: 46; Nagel 1991: 65–76). Its biggest proponent, the Reich Leader of the SS, established in the Ahnenerbe an independent research department with affiliated observatories, in order to prove its correctness. Under the camouflage of 'Institute of Meteorology' (Pflegstätte für Wetterkunde) astro-nomers, geologists, mineralogists and meteorologists worked on its empirical substantiation (Kater 1974/1997: 48, 51–3, 86–7, 97). Himmler was personally concerned with the dissemination of relevant literature and attempted to silence opponents of the *Welteislehre*. These opponents were largely to be found in the ranks of German academia, which, according to him, unfortunately showed 'a regrettable back-wardness in research and knowledge' (Himmler to Heydrich, 14 August 1938, cit. in Ackermann 1970: 46–7; see also Nagel 1991: 80–2, 85–7).

Above all the *Welteislehre* appears to be a particularly grotesque example of one of those ideologically distorted, totalitarian pseudo-sciences, by which the relationship of National Socialism to the sciences has been characterized for a long time. In spectacular contradiction to the secure knowledge of physics and astrophysics, it was based on four wholly untenable fundamental assumptions: on the existence of cosmic ice as the building material of the world; on the assumption of hot giant stars with masses millions of times greater than the sun; on the illusion that the cosmos was filled with ether which curbed the movements of the heavenly bodies; finally on the assertion that, instead of Newton's Laws, other cosmic ballistics applied in outer space (Nagel 1991: 18). Moreover, the all-embracing 'glacial-cosmogony' explanation for the origins of the world was charged with *völkisch* racial ideologies. Its National Socialist proponents recognized in it 'a unified, fantastic and splendid world picture, a scientific foundation for a truly Nordic *Weltanschauung*', a powerful German answer to Einstein's 'dreadful and mistaken theory' and its 'flooding waves of spiritual destruction' (cit. after Nagel 1991: 62, 60). In the cosmic struggle between ice and heat they saw the realization of a 'basic nordic, heroic standpoint'. To verify the *Welteislehre* empirically against the resistance of established science, and to gain acceptance and popularity for it, was the fundamental task of the Ahnenerbe. In the Ahnenerbe as the 'living armourer' of the *völkisch* spirit, the 'territory, spirit, deeds and inheritance of the northern racial Indo-Germans' was to be researched through the application of scientific methods (cit. in Ackermann 1970: 44).

The enthusiasm of the SS for researching what were, in their opinion, scientifically neglected, but ideologically important areas is itself instructive. At a very early stage already, Himmler used the opportunities offered to him by the expanding SS-Empire to build up new institutions to counter the existing scientific system. In addition to the Ahnenerbe, to mention only one example, there was the Sicherheitsdienst (SD) of the SS, where, in 1935, the Reich Leader established a large scale academic project for the research of the persecution of witches. The so-called 'H-Special Commission' (H-Sonderauftrag) later became an independent department of the Reich Central Security Office, where in 1941 it had established itself in Office VII (research and evaluation of the world view), and, more precisely, in Referat C3 for 'academic special commissions'. Himmler's concern was to prove through a quantitative recording of witch trials in the Old Reich that the persecution of witches had been a crime of the Catholic Church and the Jews against 'healthy Germandom', and to uncover the rest of the ancient German, popular pagan culture. This work continued until 1944, resulting in a 'witch card index' with around 33,000 index cards from over 260 archives and libraries (see Lorenz et al. 1999; Hachmeister 1998).

Himmler in no way limited himself to Germanic culture, Aryan linguistics, prehistory and early history, and so on. He was also concerned, as the Ahnenerbe research institute demonstrates, with the natural sciences. From 1937, there was a substantial increase in the creation of departments for the natural sciences; alongside the *Welteislehre*, in its disguise as 'meteorology', were, for example,

ethnology, ethno-archaeology and speleology, and medical research.
This corresponded to the organic holistic ideology of National
Socialism, which also appealed for the rejection of any 'specialization',
for the abolition of the 'liberalist' division between natural sciences
and arts, and for the opening of the sciences to the *Weltanschauung*.
The glacial cosmogony thereby gained a particular significance,
because Himmler rejected the Darwinist theory of evolution for the
Aryans, presenting instead phantasies, according to which their
earthly existence was derived from living shoots conserved in the ice
of outer space (Kater 1974/1997: 50, 85).

So, once more: does the *Welteislehre* represent a typical 'Nazi science'?
Here the point is not that – like social Darwinism, racism and eugenics
– it was not a product of the Third Reich, but of the turn of the century.
Since 1894, the Austrian engineer Hanns Hörbiger (1860–1931) had
brokered his doctrine in the manner of seer and prophet for a throng
of initiates. This movement organized itself in the 1920s into unions
and institutions, where it developed lively publication activity and
eventually integrated the glacial cosmogony into the *völkisch*-racist
Weltanschauung (Nagel 1991: 29–63). Instead, the starting point for
these reflections is that until well into the 1970s contemporary history
and the history of science started from the assumption that in National
Socialism they were dealing fundamentally with an anti-scientific
movement. The Third Reich was supposed to have promoted dilettante and ideologically distorted pseudo-sciences, while highly
qualified scholars were hindered, gagged or driven out on ideological-
political grounds; the twelve years of brown rule were said to have
misused, severely damaged and finally destroyed science.

There are, it is true, numerous examples of anti-scientific remarks
and even actions on the part of leading National Socialists. The
völkisch-racist ideology is irrational and interspersed with myths, and
the *Welteislehre* displays all the marks of a pseudo-science. The mass
dismissals and expulsions of racially and politically 'undesirable'
scientists from universities and research institutes occupied such a
high position on the scale of priorities of the brown rulers that only
a short time after the seizure of power they introduced the Law for
the Reconstitution of the Professional Civil Service (Gesetz zur
Wiederherstellung des Berufsbeamtentums) of 7 April 1933. Around
15 per cent of qualified university lecturers with a *Habilitation* (the
expanded second doctoral thesis required in Germany in order to get
a professorship) that means 1,100 to 1,500 people, were affected. In
addition there were scholars and scientists at institutions outside the

universities, and further, the university assistants and researchers without a *Habilitation*, in all at least 2,000 additional people. Those dismissed from industrial research are not included in these figures (differentiated figures from new research are summarized in Ash 2000). Among the scientists forced to emigrate were, as it has long been argued, numerous future Nobel Prize Winners and exceptional scholars in exactly those disciplines with a promising future, which after the Second World War would make the United States into the leading scientific nation. By contrast, German universities promoted ideologues like Hans F. K. Günther, who received a chair for racial science. High-profile theoretical physicists like Werner Heisenberg were hindered in their careers, while representatives of the anti-Semitic Aryan physics occupied academic positions. With regard to the students' education this caused great scientific damage, which would still be noticeable decades later in the Federal Republic (see Wengenroth 1993a and 1993b).

This picture of ideologically deformed scientists, or even pseudo-scientists, in the Third Reich enlarged and supplemented itself with those scientists who were directly bound to the extermination machinery of the Nazis: scientists who carried out their perverted research through cruel human experiments in the concentration camps, through their participation in the selection of ill and mentally disabled people, or through medical examinations of the bodies which had been ordered from their place of death. In fact the Ahnenerbe also housed an 'Institute for Specific Research in Military Science' (Institut für wehrwissenschaftliche Zweckforschung), which accommodated projects of this type, which scorned and destroyed human beings (Kater 1974/1997: 227–64; Schleiermacher 1988). Here in the Ahnenerbe, with its spectrum ranging from dilettante to murderous natural sciences, according to the historian of science Herbert Mehrtens, '"fascist" science in its narrowest definition' was to be found (1980: 29).

Nevertheless, the theme of science and National Socialism does not end here. Looked at it the other way round, the image of dilettante, distorted or abused sciences implied that only 'bad' scientists were involved in the Nazi state, 'bad' scientists whose scientific competence had been ideologically distorted, who had abandoned scientific standards, offering themselves to an irrational ideology which could be described as deeply hostile towards science. A true Nazi could therefore never be a good scientist, because 'good' science, i.e. science committed to pure scientific principles, was synonymous with 'absence of ideology'. According to this view, National Socialism was necessarily linked with 'bad' or even pseudo-science and the destruction of 'good' science. Several examples seemed to prove this hypothesis. Apart from the largely grotesque example of glacial cosmogony, let me mention again the Aryan Physics movement which was strongly promoted by the state. This group of apparently 'bad' scientists denounced Einstein's theory of relativity as a Jewish intrigue and rejected modern physics as a whole. Although they were led by two former Nobel Prize winners, Philipp Lenard and Johannes Stark, their scientific competence had obviously been distorted by their anti-Semitic delusions (see Beyerchen 1977; Richter 1980).

If we stay with this example, however, what is to be made of the many physicists in the Third Reich who did not follow the Aryan Physics, but remained true to the scientific norms of their discipline? After 1945 references were made to the fact that 'for the physicists of Germany, even during the war the will never slackened to do really good physics and not to allow the entanglements of personal fate to defile the purity of scientific knowledge (*die Lauterkeit wissenschaftlichen Erkennens*)' (FIAT Review of German Science, vol. 14: 194, cit. in Mehrtens 1994a: 329). This quotation is typical of the self-image shared by most scientists in Germany. They regarded themselves as 'good' scientists, free of ideological distortions, honouring scientific standards and carrying out serious research. They asserted that they had not allowed themselves to become political instruments. So they could have nothing to do with National Socialism, which meant, according to this view, a deep break in the history of science. Similar patterns of argument are often to be found, for example, in a statement made by the German Physics Society in 1946. The Society published an account of its permanent struggle against Aryan Physics during the Third Reich, maintaining that it had always been a spokesman for 'pure and decent scientific physics' ('die Sache einer sauberen und anständigen wissenschaftlichen Physik') against the claims of party and state (cit. in Mehrtens 1994b: 17).

This rhetoric of 'pure', 'decent' or 'true' science, defending scientific values against politics and ideology conferred on any methodically correct research a moral value (see repeatedly Mehrtens 1980: 33–6; 1994a: 329–31; 1994b: 17). This had several consequences: in some cases simple adherence to professional scientific values was declared to be in opposition to National Socialism (see e.g. Beyerchen 1977, esp. 206–7). The argument could, however, be turned right round to legitimize cooperation with the Nazi rulers as long as it served 'true' scientific purposes. Thirdly, in this manner it was very easy to identify those relatively few 'bad' or pseudo-scientists who had given up scientific values for ideological and political reasons. This had the clear effect of absolving the majority of scientists, for until the 1960s and even the 1970s, it was mainly those ideologically distorted disciplines or criminal individuals on which historical research focused when – if at all – dealing with the relationship between science, technology and National Socialism.

It was, nevertheless, not always only natural scientists, who, supported by an appropriate self-image, deflected attention in an apologetic manner from their own entanglements in the system, when they, for example, dismissed racial research in the Third Reich as the stuff of political fanatics and marginal pseudo-scientists (see e.g. Massin 1999, esp. 12–13, for anthropology and human genetics). The contemporary self-perception of German scientists found a parallel in the explanations of contemporary historians and historians of science, which were fed by two trains of thought. One root was the theory of totalitarianism, which long dominated the political inter-pretation of National Socialism and Stalinism. Without wishing to elaborate on content and differences here, I simply deal with the assumptions shared by all theoreticians, that totalitarian regimes were a modern phenomenon and therefore underpinned their compre-hensive claims to validity with a scientific facade. The research that was carried out claimed to be objective, empirically safeguarded and methodologically correct, yet its discoveries were essentially pre-determined by a *Weltanschauung*. Science in totalitarian systems, according to this view, was completely politicized and degenerated into a mere instrument of ideology. Outcomes of scientific enquiry which stood in the way of the desired results were twisted, distorted, ignored or prohibited. Examples were provided by Lyssenko, who determined the direction of Soviet biological research in the time of Stalin, and also by the grotesque or even perverse research of Nazi scientists (Kater 1974/97: 47–50, 358–60).

A second root of this interpretation, connected to the theory of totalitarianism, that came to the attention of American historians (see e.g. Proctor 1988: 5–6, 339/fn. 7; Macrakis 1993: 6) was the early Anglo-American sociology of science. It stemmed from a supposed interdependence between political systems and scientific efficiency and consequently also came to the conclusion that the politicization of science inevitably meant its destruction. Robert Merton emphasized, for example, that science defined itself through 'a set of characteristic methods' and a 'stock of accumulated knowledge', but was regulated above all through a 'set of cultural values and mores governing the activities termed scientific'. According to his view, these imperatives of research corresponded to the main principles of western-style liberal democracy, whilst the irrationalism and the monopoly of power of totalitarian societies limited the scope of scientific creativity (1942, esp. 268–73, 278). On the side of the left, John Desmond Bernal lamented in 1939 that National Socialist rule fundamentally threatened 'the progress and even the very existence of science', because 'attempting to maintain the standard of objectivity [was] anathema to the Nazi State'. It used science for its own ends, in order, for example, to wage war or to underpin the myth of racial superiority with the means of science. Indeed the result was that 'the whole of biology, psychology, and the social sciences need[ed] to be distorted for this purpose. In the latter cases the twisting of accepted facts has to be on such a grand scale as to destroy science completely' (Bernal 1939: 210–21, quot. 211, 216, 219).

However, the situation among historians of science has changed since the beginning of the 1980s. The discussion in Germany started with a volume edited by Herbert Mehrtens and Steffen Richter (1980), who pointed out that it had been the struggle for the alleged autonomy of 'pure science' which led scientists to collaborate with the Nazis (Mehrtens 1980: 66). The traditional view of the relationship between science, technology and National Socialism was radically questioned by a group of largely younger German historians of science working in close contact with their American, French and British colleagues. Several edited volumes emerged from this new critical approach to the history of science which surveyed the results of various research projects (Olff-Nathan 1993; Renneberg/Walker 1994b; Meinel/Voswinckel 1994). They corresponded to similar efforts by general historians who were trying to achieve a critical history of the arts and humanities (Lundgreen 1985). The former construction of a contrast between 'bad', distorted science and 'good', true science, being highly ideological itself, gave way to a rational, neutral 'Science is what scientists do' (Proctor 1988: 9). Therefore, the German racial hygienists, for example, could be seen for the first time as an integral part of the international eugenics movement, which claimed to draw up, using the natural sciences, a hierarchy of more or less 'valuable' people and to work for 'racial improvements' through positive as well as negative measures (see e.g. Kühl 1994 and 1997). The focus of historical research broadened considerably. It no longer reduced science to basic research, and it included for the first time not only the leading representatives of their disciplines, but also the students (Grüttner 1995) and the majority of scientists doing applied research or working in industrial laboratories, who had previously been completely ignored. Scientific and university history publications since then have reached an astonishingly high number (see e.g. the bibliographies in AStA 1998 and 1999), even when reviewers lament the prevalence of quantity at the cost of quality in some areas (Jansen 1998). Fundamentally, however, a relapse into all kinds of simplistic black and white representations is no longer possible, making the way clear for numerous differentiated analyses of the very complex relationship between the sciences and National Socialism (see e.g. K. Fischer 2000).

II

I would like to mention just a few important considerations and consequences resulting from this new beginning.

First, contemporary historians and historians of science now reflect more strongly the possible point of reference of their research (see Lundgreen 1994: 117–19). If one selects, as used to be common practice, National Socialism as a political system as a point of reference, then the examination of the politics of universities and of science in the Third Reich means a further contribution to the research on the Nazi regime. From this point of view the year 1933 emerged as a break, marking the beginning of a comprehensive policy of *Gleichschaltung* for universities and research institutes: the removal of the autonomy of universities and the introduction of the *Führerprinzip*; the mass dismissals of politically and racially 'undesirable' scientists and their forced exile; the influence of the state and party organization over appointments, teaching and research; the centralizing measures in the course of the *Gleichschaltung* of the *Länder* etc. From this point of view the regime was proactive, while the scientific system was predominantly reactive, or even manipulated. The frequent use of the 'word "abuse" of science with its echoes of sexual abuse and violated innocence' proves, however, to be inappropriate (Mehrtens 1994b: 24).

Secondly, if one places science as a cognitive and social system in the centre of historical interest and focuses one's attention on the

behaviour, writing, speeches, deeds, and in particular on the research
of scientists, something that has been done increasingly since the
1980s, the caesura character of the National Socialist seizure of power
is less apparent. The way is therefore cleared for attention to be
turned to the important issue not only of the breaks, but also of the
continuities – in personnel, content and institutions – in the history
of German science and universities, and indeed as much before 1933
as after the Second World War. With this change in perspective
scientists appear as proactive subjects, who understood how to use
the new political conditions for their own ends. As well as compulsory
measures, the political interventions and the instrumentalization of
science, numerous examples of the self-*Gleichschaltung* of scientists
and their offering their services for political and military ends stand
out. Empirical examinations of individual universities and research
institutes have developed a differentiated spectrum of different types
of self-*Gleichschaltung*, behind which the examples of disciplinary and
institutional self-assertion clearly receded (see, for Tübingen, Langewiesche 1997). In his research, Mark Walker has interpreted the
relationship between physics and National Socialism as 'one of
compromise and collaboration' and on top of that called for all
disciplines to examine the 'grey areas', where scientists 'both opposed
and supported certain aspects of National Socialist policy' (1989b:
85; see also Walker 1989a). For example, the ambivalent position of
Max Planck as President of the Kaiser Wilhelm Society and also as
Secretary of the Prussian Academy of Science has been analysed with
greater differentiation and more critically than previously (Albrecht
1993; Walker 1995: 65–93; Hoffmann 2000).

   If one considers individual people instead of whole disciplines, it
emerges that for geographers, for example, the year 1933 meant neither
a thematic nor a personnel break. The scientific concept of *Lebensraum*
('living space'), which became a political and military programme in
the Third Reich, had already been developed at the turn of the century
by a German geographer (Friedrich Ratzel 1901). Since then it had
also influenced other cultural-scientific disciplines. Politically, in 1933
the majority of geographers were very sympathetic to territorial
expansion into the *Lebensraum* claimed by Germany. They actively
made the most of the possibilities offered to them by the new regime
and received new research institutes for their subject, support for
projects, financial and personnel resources, and not least social
recognition (Rössler 1990; see also Rössler in this volume). For the
period of the Second World War from 1942, mathematics (Mehrtens

1996: esp. 113–18) and above all aviation and aerospace research (Trischler 1994 and 1992: 241–83) have proved that scientists mobilized themselves for the war effort in considerable numbers (the term 'self-mobilization' (*Selbstmobilisierung*) was used for the first time by Ludwig 1974: 251). Therefore, the corrupted or abused scientists now became participants, who, 'growing from within science itself', had a role in the construction of national socialist politics (see Proctor 1988: esp. 3–5, quot. 4).

Thirdly, recent critical history of science has fruitfully taken up the results of research in contemporary history, which emphasizes the polycratic character of the Nazi regime, its institutional anarchy and permanent internal power struggles. The decision to abandon the model of a monolithic governing block provoked the thought that the relationship between scientists and National Socialism was neither as a one-sided relationship dominated by the regime, nor a simple two-way relationship. It had far more to do with a many-sided, exceedingly complicated interplay between numerous participants. The individuals, institutions and organizations of science were interwoven in a network of relationships. This was true as much among scientists themselves as among the different power blocks such as the NSDAP, the SS, the state bureaucracies, the military, big industry etc. The ability to cooperate and to compete was decisive in opening up opportunities to scientists. This complex cooperation and conflict was characterized by pressure and force, but also by an unexpected scope for manoeuvre. Scientists could be constrained by the close entanglements and also the permanent disruptions created by the internal power structure, but these changes also provided them with considerable space for creativity and freedom of action, if they found the right partner in the power poker game (Renneberg/Walker 1994a: esp. 1–4). Numerous examples from the Third Reich confirm above all the increasing importance of individuals, their ambitions and miscalculations in the network of permanent rivalries (see for example Walker 1995: 5–63; Weiss 1996; Hammerstein 1999: 175–203).

Fourthly, the historical analysis of rival power blocks within the NS system opened the way to recognizing that the National Socialist regime was in no way simply hostile to science. Reflecting the thesis of the 'reactionary modernism' of the political right since the Weimar period (Herf 1984), the great importance of technocrats within the Nazi elite, who recognized the potential of scientific research for the National Socialist system, emerged increasingly clearly. Their influence

grew after the accelerated rearmament from 1936 and above all
following the failure of the *Blitzkrieg*, as from 1942 on they positively
grasped the potential of the self-mobilization of scientists and
understood how to use it (for aviation research see Trischler 1992:
241–83).

Contrary to previous belief, that means that it was not the researchers of runes, the proponents of the *Welteislehre* and other sectarians
who helped support the National Socialist extermination machinery,
and maintain it in operation (see Schöttler 1999: 40). As has been
demonstrated, scientists working in the Third Reich were predominantly not ideologically dogmatic, but largely rational and
results-orientated. In the end, the Aryan physics of someone like
Johannes Stark were no longer supported by the regime, because, as
Mark Walker described it, 'his science was not "Nazi science"' (1995:
63). Using the example of geography by focusing on the *Ostforschung*
(see Burleigh 1988), let us consider the so-called Generalplan Ost
('General Plan East'). From 1940 until 1942/3 ambitious plans were
developed to settle German or ethnic German colonists in Eastern
Europe. This implied as a first step the mass deportation or the
enslaving of millions of Slavs and the murder of the Jewish population.
Recent historical research has impressively underlined the great
significance of 'reputable' scientists and technocratic elites in
developing such schemes. It was not only SS fanatics who took part
in these monstrous plans, but also scientifically trained agricultural
and administrative experts, architects, demographers, geographers,
and among these not the ideologically sympathetic geopolitical
scientists, but the 'modern' space researchers, urban and regional
planners, etc. (see e.g. Aly/Heim 1991; Kahrs 1992; Rössler/Schleier-
macher 1993; Madajczyk 1994; Hartenstein 1998). The Third Reich
was based on the union of irrational ideology with the dynamic of
science and technology, so that in the inferno of its decline 'technocracy – and with it science and engineering – was emerging as one of
the most powerful and last pillars of the National Socialist state'
(Renneberg/Walker 1994a: 4–9, quot. 9).

Fifthly, the relationship of the Nazi rulers to science was not simply
hostile, but utilitarian: what was useful was good. The only irrefutable
ideological condition was the expulsion of the Jewish scientists, which
was primarily motivated not by science politics but by racial policies
(see Ash 2000). The utilitarianism of the National Socialists went
beyond Alan Beyerchen's supposition, whereby the prevailing image
of the natural sciences fundamentally corresponded to the unpolitical

self-image of the professional scientists, namely that science was something other than politics. Rather, from a Nazi perspective there were 'useful' disciplines like biology, chemistry, geography or engineering sciences, and other subjects like mathematics or physics, which had to prove their usefulness and were therefore politically vulnerable in the first instance (Renneberg/Walker 1994a: 9–11). Therefore, the theoretical sciences, which were not directly politically applicable, generally had a harder time than applied disciplines. Once a potential use was made clear, however, the rulers allowed 'useless' competitors like the Aryan Physics, völkisch cultural research or the *Welteislehre* to sink into oblivion, in spite of their ideological conformity. The SS itself ceased to promote the *Welteislehre* after its disciples had promised long-term weather forecasts, but clearly failed during the war. The Ahnenerbe as a scientific research institute proved to be a failure even from the point of view of National Socialism (Kater 1974/1997: 86–7, 222–3; Nagel 1991: 76–83; Nagel 1994: 171).

On the other hand, disciplines which had been mistakenly rated as utterly incompatible with National Socialism definitely received their opportunity, after they had proved their usefulness. So, to take some examples from the humanities, the question of the history of psychology and sociology in the Third Reich was not posed until the 1980s. Since then, however, the critical history of science has made it clear that the regime strongly promoted these disciplines after the prerequisite dismissal of all Jewish and leftist scholars, and the closing of their institutes. Psychology and psychotherapy now traded under the new label of a 'German *Seelenkunde* and *Seelenheilkunde*', where even psychoanalysis found its place, although without its conceptual framework and terminology. Generally speaking, psychologists who remained in their positions managed in different ways to prove the political suitability of their discipline, while on the other hand the regime became convinced that this profession could above all be useful in the military sphere. Psychology, still hardly established as a discipline and profession, actively used the opportunities offered by the regime. As a result of their considerable effort to become professional in the Third Reich, clear career opportunities opened up for psychologists outside the universities. Their training paths and careers were institutionalized for the first time, and their qualifications officially recognized (see Geuter 1984; Cocks 1985/1997; Ash/Geuter 1985: 72–7, 133–41, 160–8, 172–95). The situation was similar for sociology, which likewise went through a process of professionalization and differentiation mainly of its empirically orientated component disciplines. They established themselves, so to speak, as 'social space research' (Klingemann 1996: 119), i.e. as an applied and politically useful special science for foreign and border studies, agricultural and migration research, and also in the sphere of employment studies and research on marginal groups. They enjoyed political recognition as never before (see e.g. Kaesler 1984; Rammstedt 1986; Klingemann 1987; Derks 1999).

Sixthly, previous examples have made it clear that recent historical research has discovered the existence of so-called 'normal' science in the Third Reich, if you are willing to accept the term 'normal' for a situation in which all Jewish scientists had been forced to leave their positions. In spite of the forced mass emigration, the Third Reich was no 'scientific wasteland' (Macrakis 1993: 3–4). Numerous examples show that many disciplines and institutions suffered, but that in certain fields modern scientific research that maintained all scientific standards was not only possible but even promoted by the National Socialist rulers. Disciplines like area research and regional planning, which were strongly promoted by the state, fulfilled all the scientific standards of their time and continued to function without a break in personnel or methodology after 1945 (see Rössler 1990; Messerschmidt 1991; Esch 1992). Psychotherapy 'as an institutional and professional entity' experienced 'enough exposure, organization, training, and practice for the nascent profession to propel it into autonomy' after the end of the war (Cocks 1985: 4, 231). The scientific explanations offered by doctors, psychiatrists and psychologists for the causes of homosexual behaviour were fundamentally the same during National Socialism as before 1933 and after 1945; they distinguished between inborn and acquired forms of homosexuality. Sometimes the debate in the Third Reich even swung towards the last assumption and therefore to the possibility of therapy for homosexuality (Oosterhuis 1997).

After its 'self-*Gleichschaltung*', by which it accommodated itself to the regime (Glum 1964: 443), the Kaiser Wilhelm Society pushed ahead in its privileged niche with 'science as usual' (Albrecht/ Hermann 1990; for recent research see Kaufmann 2000). It flourished with a research budget which was doubled until 1939 and reached unimagined heights during the Second World War (Macrakis 1993: esp. 5, 157). In its institutes biological and medical, as well as biochemical research, were increasingly and generously promoted, sometimes supported by money from the Deutsche Forschungs-gemeinschaft or the Reich Research Council (Reichsforschungsrat).

Grants were usually based on purely scientific criteria; criteria like party membership or personal connections played no role. Predominantly, projects were promoted which represented, even in international comparison, modern, high-ranking research. In particular applied botany was supported, as were breeding research, interdisciplinary virus research, above all genetics, on which 45 per cent of the available funds were spent, as well as the closely connected experimental mutation research. In the course of a clear transfer of research from the universities to outside institutes, the Kaiser Wilhelm Society could further extend its privileged position within the German scientific landscape. However, the fields of research promoted by the regime must have appeared promising to the rulers on ideological-political grounds: the struggle against cancer, the securing of staple foods, the breeding of new resistant species of plants for the expanded German living space, the significance of genetic research for racial hygiene, etc. Names like Fritz von Wettstein, Alfred Kühn, Nikolai Timoféeff-Ressovsky, Adolf Butenandt and others speak for the high scientific quality of the projects (see e.g. Deichmann 1992/1995, 1994 and 1996; Deichmann/Müller-Hill 1994; Macrakis 1994).

The medical sphere deserves individual attention. In spite of the mass dismissal of Jewish scientists, the overwhelming majority of the medical journals continued to appear until 1938 without interruption or restriction, and more than a dozen new journals were even founded. It has been estimated that German medical journals from this time filled more than a hundred continuous metres of shelves - 'more than any other country in the world in this period' (Proctor 1988: 5). Supported by the regime, medical research in Germany achieved significant results in areas which today seem very 'progressive' and socially responsible: in the sphere of the long-term effects of environmental pollutants or the toxic effects of tobacco, heavy metals, asbestos or alcohol. Nazi doctors propagated a healthy diet rich in roughage, warned against the damaging effects of alcohol, and successfully pushed for the introduction of stricter health and safety regulations in the workplace. They were, as Robert Proctor recently confirmed, pioneers of precautions against cancer. They established proof of the connection between smoking and lung cancer, coined the term of passive smoking and led the first public anti-smoking campaign in modern history. It should be added that the motivations behind the 'Nazi War on Cancer' were the same as those for criminal medical experiments: a racially hygienic utopia reserved for healthy and ever healthier Germans but excluding the bodily and mentally ill and the *Blutsfremde* ('people of alien blood'), who were to be exterminated (Proctor 1999).

A seventh consequence of the new approach to research in the history of science since the 1980s is a differentiation and fresh assessment of the results of the emigration of scientists after the National Socialists took power (see Strauss 1991; W. Fischer 1994; Ash/Söllner 1996; Krohn et al. 1998). On the one hand, it has been shown that the dismissal and emigration of scientists defined as 'Jewish' by the National Socialists was very unevenly spread and varied in proportion to their respective contributions to the individual universities, institutes, regions and disciplines (see for mathematics Siegmund-Schultze 1998 and Deutsche Mathematiker-Vereinigung 1998; for physics K. Fischer 1988; for biology Deichmann 1992/1995 and 1996; for chemistry Deichmann 1999 and in this volume). On the other hand, the qualitative consequences of the expulsions were no longer decided simply according to 'loss and gain'. The earlier static view was given up, whereby, as in a zero sum game, the later achievements of the emigrants in the countries to which they migrated meant an equal loss to Germany. It made space for a dynamic interpretation, which no longer assumed the transfer of finished scientific projects between closed systems. Against a background of research projects on national styles of scientific thought, cultures of innovation and internationalization of knowledge, new questions are being asked today about the international circulation of elites and scientific change in open cultural systems (Ash/Söllner 1996; Ash 1999 and 2000).

The task of a critical history of science has become harder due to the results of recent research. Simplistic explanations and models have proved to be wholly insufficient in the face of the complexity of relations, and the heated discussion about modernization and National Socialism has led no further in the case of science (see e.g. Prinz/Zitelmann 1991). The debate about the recently (re-)discovered involvement of some of the most significant historians of the Federal Republic in the mobilization of the discipline of history in support of the Third Reich (Oexle/Schulze 1999; Hohls/Jarausch 2000; see also Schönwälder 1992; Schöttler 1997; Wolf 1996) was strongly centred on individuals. It raised with renewed urgency, however, the continuing question, also much discussed in the media, of the position and function of the arts in National Socialism. Besides considering individual disciplines (see Hofter 1998; Flashar 1999; Hausmann 2001), researchers concentrated on long unknown, big interdisciplinary projects like the voluntary 'deployment for war' of the arts in the 'Ritterbusch action' (Hausmann 1998). An even more specific example is the network of ethnic German research communities, which in 1940 employed about 1,000 people in different arts and social science disciplines. They had considerable financial resources at their disposal and a regionally and functionally strongly differentiated structure of organization. They functioned as an interdisciplinary brains trust for the 'Germanization' of the annexed living space in eastern and south-eastern Europe (Fahlbusch 1999).

III

The self-mobilization of human and natural scientists on the one hand and their close connection with the formulation and execution of National Socialist politics on the other, prompt reflection on the formation of society in a wider framework, in which politics and science stand in close reciprocal relationships and clearly need, use, stabilize and legitimate each other. Over and above individual empirical research, in my opinion, the sociological concept of the 'knowledge society' offers a starting point that is worth consideration (see e.g. Böhme/Stehr 1986; Stehr/Ericson 1992; Gibbons et al. 1994; Stehr 1994a and 1994b). Developed in the 1980s to explain the society of the future, this concept could be applied to history and to National Socialism as well.

Modern societies are knowledge societies. It is true that archaic societies are also based on knowledge, but here it is a matter of comprehending terminologically and conceptually the formation of an advanced society, in which scientific knowledge grows quantitatively in an exponential manner and above all continually broadens its social function, until it penetrates all public and private areas of life including the intimate sphere. In this most recent phase of social development, work and production lose their power to structure and organize society, while scientific knowledge becomes the driving force of social change. Using Marxist principles, science, having previously been a phenomenon of the social superstructure, increasingly emerges on the level of society's material base, a productive force the significance of which is growing in relation to the declining significance of other productive forces. The traditional class contrasts between capital and work are overlaid with new fields of action which follow other rationales and have new protagonists, whose function and social influence are based on knowledge. The system of science and knowledge now stands in a structurally new type of performance relationship to the subsystems of its surroundings like politics, economics or culture. The boundaries of the systems dissolve; knowledge permeates all spheres.

In the search for the trail of the emerging knowledge society, historical research has, in the recent past, focused on new thematic fields: on the precautions and measures which a society takes in order to produce and impart knowledge; on the institutions in which this occurs; and finally on the social groups which are the bearers of this knowledge. 'Knowledge' goes far beyond 'science' both conceptually and theoretically. At the core of the term we find scientific and technological knowledge, but it encompasses the humanities as well. It also places a special emphasis on interpreting, orienting and action-defining knowledge, thus comprising all that we express with the German term *Bildung*. Furthermore, it consciously opens its gates to non-scientific forms of knowledge like religious, symbolic and visual knowledge. By this inclusiveness, the concept does not mean a zero sum game in which the purportedly 'modern' elements increase continually while the extent of traditional, non-scientific knowledge is correspondingly on the decrease, unable to change or develop on its own. In any case, scientists as a social group deserve far more attention than before. Questions arise concerning their position in social structures, their connection to power structures or their function in the legitimation of government. In addition to the scientists who produce new scientific knowledge, there are the experts, who reproduce existing knowledge and thus make it generally accessible. The experts play a central role in the sociology of the knowledge society.

Experts provide access to specialist knowledge for those seeking advice on matters for which they themselves do not have information at their disposal, but of whose significance they are aware. This means that parallel to the experts there also develops an increasing clientele in government, the civil service, business, etc., who require specialist knowledge. In emerging knowledge societies, therefore, the practice of government and power is mediated to an increasing degree by experts. Further, it also means that experts do not operate passively, but function actively in their socially highly significant intermediary position between the producers and the consumers of knowledge: '[They] define the situation for the untutored, they suggest priorities, they shape the people's outlook on their life and world, and they

establish standards of judgement in different areas of expertise'
(Rueschemeyer 1986, cit. in Stehr 1994a: 360). With reference to
scientists and experts it is also worth considering that they likewise
make social demands on scientific expertise through self-induction,
in which they define socially relevant problems and fields of conflict,
and then promise to solve them with the help of their science. There
therefore exists a more or less mutually dependent relationship
between science and politics in which each legitimates and is the client
of the other.

A perspective therefore emerges which links the Third Reich in a
hitherto little noted manner with its immediately preceding and
succeeding history. The 'scientification' (*Verwissenschaftlichung*) of wider
spheres of life was already present in Germany at the turn of the
century at a level that justifies us in detecting an increasingly clear
path of a slowly emerging knowledge society since the 1880s. I am
referring for example to the countless experts in the human sciences,
who since the *Kaiserreich* were permanently and effectively present,
with their research results, in the state and communal welfare
systems, in poor relief, social security, public health, or public
transport (see Raphael 1996). In any case, the Nazi *Machtergreifung*
came at a time when progress had already been made in the process
of 'scientification' in wider social spheres. The Nazi regime definitely
intervened in this development, but with varying levels of enthusiasm.
It was destructive when ideological concerns stood in its way, but it
in no way caused the process to decline. It also transformed structures,
bound 'state and military institutions in a new way into the complex'
(Mehrtens 1994c: 250) or adapted itself to realities. The great social
significance of scientists and experts in government, civil service,
business etc., as well as their active participation in the construction
of society and politics, has to be emphasized also for the Third Reich.
Given the continuing and deepening reciprocal relationships between
science and politics, both the natural and the human sciences should
not be depicted as staying in the self-chosen isolation of an unpolitical
'garden of true science' (Wilhelm Süss in FIAT-Review of German
Science, vol. 1, cit. after Mehrtens 1996: 121).

Above all, the concept of the knowledge society presents in my
opinion a useful framework of analysis for contemporary history,
because in Germany the aforementioned 'scientification' appeared
very early compared to elsewhere, and it was also this process which
led to National Socialism. Thus chemistry, as a 'science-based
industry', confirms the growing significance from the 1880s of

scientific research for production, economic growth, intensified armament and the waging of war. Since its emergence in the nineteenth century it was directed towards the replacement of natural with scientifically developed, synthetic dyes. In the First World War, under the pressure of the shortage of raw materials and the food crisis, it conclusively followed the path of developing synthetic substitutes. In the protected war economy, innovations like synthetic nitric acid and Haber-Bosch ammonia were established in big industry (Szöllösi-Janze 1998: 270–316). In the Weimar period this trend became even stronger. There was now a general concern to realize an economy that was independent from foreign imports and therefore potentially 'capable of waging war' through the deployment of research and technology. With state support there emerged a whole system of research institutes, industrial laboratories and university institutes, in which big industry was closely bound to all levels of the production of knowledge. Thus, for example, all the Kaiser Wilhelm Institutes whose research was orientated towards the practical application of science in industry, carried out projects directed towards economic self-sufficiency, assuming the cutting off of Germany from the free world market and indirectly anticipating the next war. Therefore, in the 1920s, a unique scientific society came into being in Europe, in which in 1930 the share of the public and the private expenditure for research and development amounted to a good 1 per cent of the net domestic product. In Great Britain, which had almost twice the economic power, this was only 0.4 per cent. If one considers the scientists and engineers employed by the state and in industry, the figure for Germany was around 20,000; for Britain it is estimated at only 6,000 to 10,000, because British foreign and trade policies were directed and remained directed at the world market (Marsch 2000). Admittedly the extraordinarily high German expenditure on research and development, resulting from the policy of autarky, led the country into a dangerous technological trajectory and finally to political and economic isolation and catastrophe. Thus the management of IG Farben, which found itself facing a dead end pursuing the scientifically and technologically highly demanding but economically unprofitable hydrogenation process, was structurally susceptible to National Socialism (Hughes 1969; for biotechnology see Marschall 2000 and in this volume).

    With the collaboration of state, military, economy and science in chemistry we are dealing, in my opinion, with a highly relevant constellation for National Socialism, which is worth examining. Here members of the different societal subsystems congregate out of common interest in order to carry out large-scale scientific research for a specific purpose. Scientists had already played a substantial role in achieving long-term cooperation before, and increasingly during the First World War, in which they acted as personal mediators, organizers and innovators (see Szöllösi-Janze 2000). As mediators they furthered the connection of science with politics and big industry, and thereby strengthened the need in society for scientific expertise, paving the way for the advance of more experts from the knowledge-based professions. As organizers they endeavoured to anchor science and technology permanently in cooperative relationships, in order to institutionalize their contribution to the solution of social problems. Finally, the same technocratic patterns effectively stimulated innovations in the research sphere, which demonstrated clear similarities to modern big science. This new type of organized scientific research, which was successfully established during the First World War in some Kaiser Wilhelm Institutes and thereafter significantly intensified, crystallized, so to speak, the growing role of science, which continued into the Third Reich. The secular trends of modern research, such as the trends towards big science, towards the increased application of technology to basic research, and towards the transfer of research from the universities to institutes in the intersection of politics, economics and science, were maintained in the Third Reich. 'All the indicators of institutionally organized research basically pointed to growth, expansion, differentiation and specialization, to the scientif-ication of praxis, the close coupling of state, military or industrial needs and scientific efforts' (Lundgreen 1994: 123).

It is important to recognize and to stress that processes of 'scientif-ication' do not necessarily have anything to do with democracy. The sociology of the knowledge society defines modernization neutrally as the extension of individual and societal possibilities for action on the basis of knowledge. The German example imposes an authori-tarian and ultimately totalitarian political development through substantial inclusion of scientific expertise. The transfer of scientific rationalization from the technical to the social spheres such as working processes, housing, or health makes this particularly clear. Rationalization always means standardization, the scientific differ-entiation of standard deviations followed by the attempted institution of a norm (see Peukert 1986 for youth welfare before 1933). In the medical sphere, disciplines like psychiatry, social and racial hygiene implemented scientifically based standardizations from the turn of the century onwards, firmly establishing 'deviations' from the biological and social norm. Burdened with social-Darwinist and racist perceptions, they provided therapy for such deviations or, in the extreme case of the Third Reich, they annihilated them. It should be emphasized that we are dealing with 'medicine that, in view of its consequences, was inhuman, but hardly unscientific' (Lundgreen 1994: 124). 'Scientification' in the sense of standardization excluding deviations from the norm is a thoroughly ambivalent process. On the one hand, it expands opportunities for societal interaction in all directions. It can also, however, be inhuman, destructive and fatal.

The concept of the knowledge society does not comprehend 'scientification' ahistorically as a quasi-automatic wearing-down and self-regulating course of events. It does not suggest a linear course, but on the contrary underlines the ambivalence and tensions within the development, and also refers expressly to counter-currents and opposition. This precludes the danger of misunderstanding National Socialism as a phase in a linear progression towards the knowledge society of the twenty-first century, because the obstacles, distortions and devastation which the twelve years of Nazi rule also represented are too numerous. Even for those scientific fields supported by the regime there was the question, independent of the defeat of Germany in the Second World War, of their middle- and long-term prospects. In order to participate permanently in the exponential expansion of scientific and technological knowledge, unhindered communication among scientists was necessary, as was free access for as many as possible to this knowledge. This was increasingly difficult under National Socialism, above all during the war. Germany was not only increasingly cut off from international development. Internally, the extensive regulations concerning secrecy blocked scientific publications and also generally restricted any continuous cooperation. Particularly clear is the extreme example of the Ahnenerbe scientists, who, in spite of the cameraderie, worked in complete isolation as 'scientific autocrats' and stuck closely to their individual instructions distributed by the Reich Leader, while modern big science, like the simultaneous American atom bomb project, meant extensive team-work (Kater 1974/1997: 223–4).

The example of big science, which promoted a particularly close coordination of state, big industry and science, makes the dilemma of the Third Reich with regard to modern research and development clear. The state as the political organizing force among the societal subsystems saw itself confronted with new demands to manage and direct this interplay. Nevertheless all attempts at direct and indirect steering failed, as the examples of nuclear fission and aviation research demonstrate, until during the final phase of the war, scientists almost regained their former autonomy (Walker 1989a; Trischler 1992: 281–2). The regime proved itself in the medium term to be incapable of systematically promoting research and translating innovations into reality. Its self-destructive fragmentation into a multitude of rival agents hindered the lasting imposition of the advanced political direction of research.

IV

This volume provides not so much a survey as some glimpses at recent historical research in the complex field of science in the Third Reich. Each chapter will address a different aspect, taking as its starting point either a discipline or a technology, or the biographies of individual scientists. In general the concern is to integrate National Socialism into a long-term development and therefore to thematicize explicitly the question of continuities and breaks. The contributors do this either by looking back to the turn to the twentieth century, or by following their theme into the time of the Federal Republic.

At first sight, Sylvia Paletschek has chosen the German universities as an institutional starting point, but a university means much more than a mere institution, much more than the body of its assembled professors. She analyses the academic discourse about the idea of the German university, linked to the name of Wilhelm von Humboldt, but actually an invention of the turn of the twentieth century. Historical hindsight was used to legitimize an idealistic university concept, which was built on and conserved bourgeois educational privileges. The National Socialists formulated no self-contained programme for the universities, but directed their fragmentary attempts at overcoming impractical neo-humanist ideals. They made room for reflections on science as a useful part of politics, rather than a non-utilitarian *l'art pour l'art*. After 1945, the renunciation of National Socialism in both German states led to a renaissance of the Humboldtian idea of the university, which thereafter for decades blocked attempts at university reform.

In her investigation, Mechtild Rössler deals with a discipline: geography. Going back chronologically to the turn of the twentieth century she shows, first of all, the emergence of central geographic terms and concepts and their increasing *völkisch*-racist content. Secondly, she reconstructs the practical relevance of the geographic sub-disciplines, such as ethnic geography, area research and regional planning, colonial geography and landscape ecology, for the conquest of German *Lebensraum* during the Second World War, and she also deals with the popular SS expeditions to the Himalayas. Finally she sketches the unbroken development of area research and regional planning after 1945.

For aeronautical research, Helmuth Trischler examines the tense relationship between big science and small science in the Third Reich and highlights the problem of direction, with which the regime found itself confronted. He establishes proof that characteristics of big science first began to develop at the turn of the century, and increased during the First World War. After the barren Weimar period, the National Socialist seizure of power meant for those scientists involved in aviation the fulfilment of all their wishes: aeronautical research expanded massively in the course of military rearmament and the building up of the Luftwaffe. Helmuth Trischler reconstructs the unsuccessful direct and indirect attempts of the regime to direct research, until the power-holders eventually, at the end of the war, had to fall back on the self-mobilization and self-organization of the scientists.

Luitgard Marschall focuses on biotechnology, which stood and still stands in a very interesting tension between science, big chemistry and politics. She integrates the evolution of biotechnology in Germany into the general trends of chemical research and development since the turn of the century. With the militarily, politically and econo-mically based decision to use synthetic chemistry during the First World War and then in the Weimar Republic, the German chemical industry followed a trajectory of technological development which marginalized biotechnology. The active policy of autarky of the Third Reich strengthened this trend and granted biotechnology only subsidiary functions in certain niches during the war. By continuing her study up to the 1970s, however, Marschall demonstrates that the long-term power of this developmental pattern was independent of the regime for decades: German chemistry, and then the phar-maceutical industry, did not abandon the trajectory of chemical synthesis. When biotechnology later emerged as the innovative technology of the future, it was hardly possible to bridge the gaps in the infrastructure of research and development.

Cay-Rüdiger Prüll deals with the long-term evolution of a medical discipline in the German and British capitals, Berlin and London, since the nineteenth century. Through a systematic comparison he shows how pathology was influenced in its content and institutional development by the different political systems, health systems and scientific cultures. German pathology, which traditionally focused on corpses, was closely connected to the military, while in England it was clinically based and therefore open to the innovations in modern biochemical and bacteriological research. In the Third Reich pathologists tried to ward off the great pressure on their discipline to modernize by closing ranks with the National Socialists. This they managed to do in two ways: through the dismissal of Jewish scientists, who tended to specialize in forward-looking, modern areas, and through the marginalization of 'threatening' innovations, which they integrated into the old institutional structures of pathology, thereby hindering their independent development as innovative (sub-)disciplines in their own right.

Stefan Kühl, dealing with the example of the so-called 'euthanasia action' of 1939/40, examines a little-observed aspect of the international eugenics movement after the First World War. Unlike marriage restrictions or sterilization, the killing of disabled people had no systematic place in the logic of eugenics, as one could not promise any positive effects from it concerning the improvement of the racial base of the population. The debate that has run since the 1920s over the killing of so-called 'lives not worth living' occurred to a large degree outside the community. Starting from the 'dysgenic' effects of modern war through the huge destruction of the best and the fittest, in the 1930s the eugenicists propagated instead the declaration of an international peace order according to the motto 'race policy as peace policy'. The same racial hygienicists, however, participated in the 'T4-Killing Action'. Stefan Kühl traces the failure of this peace policy back to the outbreak of the Second World War, which unleashed the fatal logic of 'racial improvement' theories in modern war.

In his chapter on brain research, which was the contemporary expression for what we now call neuro-science, Josef Reindl has chosen a biographical approach, in order to clarify the central questions of scientific research on a micro level: which fields to enter, with which methods, and to what end? He analyses the biographies and careers of three outstanding scientists at the Kaiser Wilhelm Institute for Brain Research in Berlin-Buch. As he demonstrates, all three of them, despite being very different as far as their fields of research, methods and professional formation were concerned, used the opportunities arising from National Socialism on their own initiative. They regarded themselves as apolitical and only concerned with the advancement of 'pure' science. Their scientific vision of improving human kind through a set of biological-racial measures was in the mainstream among the scientific community of their time, but Germany was the only country in which concepts of this kind were used for mass murder.

Ute Deichmann's study links the mass dismissal and expulsion of Jewish chemists and biochemists in the early years of the Third Reich with the early Federal Republic. First, she determines the quantitative side of the exodus, which, at 26 per cent, approximately corresponds to the numbers of physicists who left, and confronts the findings with the non-reaction of so-called 'Aryan' scientists. In a second large section she assesses the correspondence of émigré chemists and biochemists with their former colleagues in Germany after 1945 and reconstructs three types of correspondence: letters from non-Nazi scientists who were reluctant to face the past; letters from compromised scientists, who now expected help; finally correspondence which shows something like normalization, understood as a process, at the end of which the National Socialist past was of only secondary importance in comparison with purely scientific matters. She finds that, with a few exceptions, no normalization of this type took place.

The essays presented here are based on papers given in Hilary Term 1999 (January until March) in my seminar 'National Socialism and the Sciences' at St Antony's College, Oxford. First, I would like to thank the contributors for their cooperation and the participants in the seminar for their profound comments in our always lively discussions. To the Stifterverband für die Deutsche Wissenschaft I owe my thanks and recognition for its financial support for the lecture series and subsequent publication, and above all for the chance to spend the academic year 1998/9 as the Stifterverband Visiting Fellow at the University of Oxford. For the hospitality and openness I received in Oxford, I want to thank St Antony's College and above all Mr Tony Nicholls and Mrs Jennifer Law, who made me feel so much at home and made it possible for me to carry out research at the European Studies Centre.

References

Ackermann, Josef (1970), *Heinrich Himmler als Ideologe*, Göttingen: Musterschmidt.

Albrecht, Helmuth (1993), '"Max Planck: Mein Besuch bei Adolf Hitler" – Anmerkungen zum Wert einer historischen Quelle', in Helmuth Albrecht (ed.), *Naturwissenschaft und Technik in der Geschichte*, Stuttgart: Verlag für Geschichte der Naturwissenschaften und der Technik, 41–63.

—— and Hermann, Armin (1990), 'Die Kaiser-Wilhelm-Gesellschaft im Dritten Reich (1933–1945)', in Rudolf Vierhaus and Bernhard vom Brocke (eds), *Forschung im Spannungsfeld von Politik und Gesellschaft. Geschichte und Struktur der Kaiser-Wilhelm-/Max-Planck-Gesellschaft aus Anlaß ihres 75jährigen Bestehens*, Stuttgart: Deutsche Verlagsanstalt, 356–406.

Aly, Götz and Heim, Susanne (1991), *Vordenker der Vernichtung. Auschwitz und die deutschen Pläne für eine neue europäische Ordnung*, Hamburg: Hoffmann und Campe.

Ash, Mitchell (1999), 'Scientific Changes in Germany 1933, 1945, 1990: Toward a Comparison', *Minerva*, 37, 329–54.

—— (2000), 'Emigration und Wissenschaftswandel als Folgen der nationalsozialistischen Wissenschaftspolitik', in Doris Kaufmann (ed.), *Geschichte der Kaiser-Wilhelm-Gesellschaft im Nationalsozialismus. Bestandsaufnahme und Perspektiven der Forschung*, Göttingen: Wallstein.

—— and Geuter, Ulfried (eds) (1985), *Geschichte der deutschen Psychologie im 20. Jahrhundert. Ein Überblick*, Opladen: Westdeutscher Verlag.

—— and Söllner, Alfons (eds) (1996), *Forced Migration and Scientific Change. Emigré German-Speaking Scientists and Scholars after 1933* (Publications of the German Historical Institute, Washington, D.C.), Cambridge: Cambridge University Press.

AStA der Universität Mannheim (ed.) (1998), *Hochschulen 1933–1945. Bibliographie* (Schriftenreihe des AStA der Universität Mannheim 4), Mannheim.

—— (ed.) (1999), *Hochschulen 1933–1945. Nachtrag zur Bibliographie sowie Übersichten über Rehabilitationen und Gedenken nach 1945* (Schriftenreihe des AStA der Universität Mannheim 6), Mannheim.

Bernal, John Desmond (1939), *The Social Function of Science*, London: Routledge.

Beyerchen, Alan D. (1977), *Scientists under Hitler. Politics and the Physics Community in the Third Reich*, New Haven: Yale University Press.

Böhme, Gernot and Stehr, Nico (eds) (1986), *The Knowledge Society. The Growing Impact of Scientific Knowledge on Social Relations* (Sociology of the Sciences 10), Dordrecht: D. Reidel.

Burleigh, Michael (1988), *Germany Turns Eastwards. A Study of Ostforschung in the Third Reich*, Cambridge: Cambridge University Press.

Cocks, Geoffrey (1985/1997), *Psychotherapy in the Third Reich. The Göring Institute*, New York/Oxford: Oxford University Press; 2nd rev. and expanded edn: New Brunswick: Transaction.

Deichmann, Ute (1992/1995), *Biologen unter Hitler. Vertreibung, Karrieren, Forschung*, Frankfurt, New York: Campus 1992; rev. edn: *Biologen unter Hitler. Porträt einer Wissenschaft im NS-Staat*, Frankfurt a. M.: Fischer

—— (1994), 'Die biologische Forschung an Universitäten und Kaiser-Wilhelm-Instituten 1933–1945', in Christoph Meinel and Peter Voswinckel (eds), *Medizin, Naturwissenschaft, Technik und Nationalsozialismus. Kontinuitäten und Diskontinuitäten*, Stuttgart: Verlag für Geschichte der Naturwissenschaften und der Technik, 100–10.

—— (1996), *Biologists under Hitler*, Cambridge: Harvard University Press (English translation of Deichmann 1992/1995).

—— (1999), 'The Expulsion of Jewish Chemists and Biochemists from Academia in Nazi Germany', *Perspectives on Science*, 7, 1–86.

—— and Müller-Hill, Benno (1994), 'Biological Research at Universities and Kaiser Wilhelm Institutes in Nazi Germany', in Monika Renneberg and Mark Walker (eds), *Science, Technology, and National Socialism*, Cambridge: Cambridge University Press, 160–83.

Derks, Hans (1999), 'Social Sciences in Germany, 1933–1945', *German History*, 17: 177–219.

Deutsche Mathematiker-Vereinigung (1998), *Terror and Exile. Persecution and Expulsion of Mathematicians from Berlin between 1933 and 1945*. An Exhibition on the Occasion of the International Congress of Mathematicians, Technische Universität Berlin, August 19 to 27, 1998.

Esch, Michael G. (1992), '"Ohne Rücksicht auf historisch Gewordenes". Raumplanung und Raumordnung im besetzten Polen 1939–1944', in Horst Kahrs et al. (1992), *Modelle für ein deutsches Europa. Ökonomie und Herrschaft im Großwirtschaftsraum* (Beiträge zur nationalsozialistischen Gesundheits- und Sozialpolitik 10), Berlin: Rotbuch-Verlag, 77–123.

Fahlbusch, Michael (1999), *Wissenschaft im Dienst der nationalsozialistischen Politik? Die Volksdeutschen Forschungsgemeinschaften von 1931 bis 1945*, Baden-Baden: Nomos.

Fischer, Klaus (1988), 'Der quantitative Beitrag der nach 1933 emigrierten Naturwissenschaftler zur deutschsprachigen physikalischen Forschung', *Berichte zur Wissenschaftsgeschichte*, 11: 83–104.

—— (2000), 'Repression und Privilegierung. Wissenschaftspolitik im Dritten Reich', in Dietrich Beyrau (ed.), *Im Dschungel der Macht. Intellektuelle Professionen unter Stalin und Hitler*, Göttingen: Vandenhoeck & Ruprecht, 170–94.

Fischer, Wolfram (ed.) (1994), *Exodus von Wissenschaften aus Berlin. Fragestellungen – Ergebnisse – Desiderate. Entwicklung vor und nach 1933* (Forschungsbericht/Akademie der Wissenschaften zu Berlin 7), Berlin/New York: de Gruyter.

Flashar, Martin (1999), 'Trübe Vorgeschichte. Ein Disput über deutsche Prähistoriker im Nationalsozialismus', *Frankfurter Allgemeine Zeitung*, 190, 18.8.1999: N 6.

Geuter, Ulfried (1984), *Die Professionalisierung der deutschen Psychologie im Nationalsozialismus*, Frankfurt a. M.: Suhrkamp.

Gibbons, Michael et al. (1994), *The New Production of Knowledge. The Dynamics of Science and Research in Contemporary Societies*, London: Sage.

Glum, Friedrich (1964), *Zwischen Wissenschaft, Wirtschaft und Politik. Erlebtes und Erdachtes in vier Reichen*, Bonn: Bouvier.

Grüttner, Michael (1995), *Studenten im Dritten Reich*, Paderborn: Schöningh.

Hachmeister, Lutz (1998), *Der Gegnerforscher. Die Karriere des SS-Führers Franz Alfred Six*, München: C. H. Beck.

Hammerstein, Notker (1999), *Die Deutsche Forschungsgemeinschaft in der Weimarer Republik und im Dritten Reich. Wissenschaftspolitik in Republik und Diktatur 1920–1945*, München: C. H. Beck.

Hartenstein, Michael A. (1998), *Neue Dorflandschaften. Nationalsozialistische Siedlungsplanung in den 'eingegliederten Ostgebieten' 1939 bis 1944 unter besonderer Berücksichtigung der Dorfplanung* (Wissenschaftliche Schriftenreihe Geschichte 6), Berlin: Köster.

Hausmann, Frank-Rutger (1998), *Deutsche Geisteswissenschaft im Zweiten Weltkrieg. Die Aktion Ritterbusch (1940–1945)*, Dresden/München: Dresden University Press.

—— (ed.) (2001), *Die Rolle der Geisteswissenschaften im Dritten Reich* (Schriften des Historischen Kolleg, Kolloquien 53), München: Oldenbourg.

Herf, Jeffrey (1984), *Reactionary Modernism. Technology, Culture, and Politics in Weimar and the Third Reich*, Cambridge: Cambridge University Press.

Hoffmann, Dieter (2000), 'Das Verhältnis der Akademie zu Republik und Diktatur: Max Planck als Sekretar', in Wolfram Fischer (ed.), *Die Preußische Akademie der Wissenschaften zu Berlin in Krieg und Frieden, in Republik und Diktatur 1914–1945*, Berlin: Akademie-Verlag.

Hofter, Mathias René (1998), 'Aus der nordisch-makedonischen Pfropfkultur. Abenteuerlicher Opportunismus: Die Altertumswissenschaft unter dem Nationalsozialismus', *Frankfurter Allgemeine Zeitung*, 262, 11.11.1998: N 6.

Hohls, Rüdiger and Jarausch, Konrad H. (eds) (2000), *Versäumte Fragen deutscher Historiker im Schatten des Nationalsozialismus*, Stuttgart: Deutsche Verlagsanstalt.

Hughes, Thomas (1969), 'Technological Momentum in History: Hydrogenation in Germany 1898–1933', *Past and Present*, 44, 106–32.

Jansen, Christian (1998), 'Mehr Masse als Klasse – mehr Dokumentation denn Analyse. Neuere Literatur zur Lage der Studierenden in Deutschland und Österreich in der ersten Hälfte des 20. Jahrhunderts', *Neue Politische Literatur*, 43: 398–440.

Kaesler, Dirk (1984), *Die frühe deutsche Soziologie 1909 bis 1934 und ihre Entstehungs-Milieus*, Opladen: Westdeutscher Verlag.

Kahrs, Horst et al. (1992), *Modelle für ein deutsches Europa. Ökonomie und Herrschaft im Großwirtschaftsraum* (Beiträge zur nationalsozialistischen Gesundheits- und Sozialpolitik 10), Berlin: Rotbuch-Verlag.

Kater, Michael H. (1974/1997), *Das 'Ahnenerbe' der SS 1935–1945. Ein Beitrag zur Kulturpolitik des Dritten Reiches* (Studien zur Zeitgeschichte 6), 1st edn 1974, 2nd rev. edn München: Oldenbourg.

Kaufmann, Doris (ed.) (2000), *Geschichte der Kaiser-Wilhelm-Gesellschaft im Nationalsozialismus. Bestandsaufnahme und Perspektiven der Forschung*, Göttingen: Wallstein.

Klingemann, Carsten (ed.) (1987), *Rassenmythos und Sozialwissenschaften in Deutschland. Ein verdrängtes Kapitel sozialwissenschaftlicher Wirkungsgeschichte* (Beiträge zur sozialwissenschaftlichen Forschung 85), Opladen: Westdeutscher Verlag.

—— (1996), *Soziologie im Dritten Reich*, Baden-Baden: Nomos.

Krohn, Claus-Dieter et al. (eds) (1998), *Handbuch der deutschsprachigen Emigration 1933–1945*, Darmstadt: Wissenschaftliche Buchgesellschaft.

Kühl, Stefan (1994), *The Nazi Connection. Eugenics, American Racism and German National Socialism*, Oxford/New York: Oxford University Press.

—— (1997), *Die Internationale der Rassisten. Aufstieg und Niedergang der internationalen Bewegung für Eugenik und Rassenhygiene im 20. Jahrhundert*, Frankfurt a. M./New York: Campus.

Langewiesche, Dieter (1997), 'Die Universität Tübingen in der Zeit des Nationalsozialismus: Formen der Selbstgleichschaltung und Selbstbehauptung', *Geschichte und Gesellschaft*, 23: 618–46.

Lorenz, Sönke et al. (eds) (1999), *Himmlers Hexenkartothek. Das Interesse des Nationalsozialismus an der Hexenverfolgung* (Hexenforschung 4), Bielefeld: Verlag für Regionalgeschichte.

Ludwig, Karl-Heinz (1974), *Technik und Ingenieure im Dritten Reich*, Düsseldorf: Droste.

Lundgreen, Peter (ed.) (1985), *Wissenschaft im Dritten Reich*, Frankfurt a. M.: Suhrkamp.

—— (1994), 'Staatliche hochschulfreie Forschung in Berlin und die NS-Wissenschaftspolitik', in Wolfram Fischer (ed.), *Exodus von Wissenschaften aus Berlin. Fragestellungen – Ergebnisse – Desiderate. Entwicklung vor und nach 1933* (Forschungsbericht/Akademie der Wissenschaften zu Berlin 7), Berlin/New York: de Gruyter, 116–26.

Macrakis, Kristie (1993), *Surviving the Swastika. Scientific Research in Nazi Germany*, New York/Oxford: Oxford University Press.

—— (1994), 'The Ideological Origins of Institutes at the Kaiser Wilhelm Gesellschaft in National Socialist Germany', in Monika Renneberg and Mark Walker (eds), *Science, Technology, and National Socialism*, Cambridge: Cambridge University Press, 139–59.

Madajczyk, Czeslaw (ed.) (1994), *Vom Generalplan Ost zum Generalsiedlungsplan. Dokumente* (Einzelveröffentlichungen der Historischen Kommission zu Berlin 80), München: K. G. Saur.

Marsch, Ulrich (2000), *Zwischen Wissenschaft und Wirtschaft. Industrieforschung in Deutschland und Großbritannien 1880–1936* (Veröffentlichungen des Deutschen Historischen Instituts London 47), Paderborn: Schöningh.

Marschall, Luitgard (2000), *Im Schatten der chemischen Synthese. Industrielle Biotechnologie in Deutschland (1900–1970)*, Frankfurt a. M./New York: Campus.

Massin, Benoît (1999), 'Anthropologie und Humangenetik im Nationalsozialismus oder: Wie schreiben deutschen Wissenschaftler ihre eigene Wissenschaftsgeschichte?', in Heidrun Kaupen-Haas and Christian Saller (eds), *Wissenschaftlicher Rassismus. Analysen einer Kontinuität in den Human- und Naturwissenschaften*, Frankfurt a. M./New York: Campus, 12–64.

Mehrtens, Herbert (1980), 'Das "Dritte Reich" in der Naturwissen-
schaftsgeschichte: Literaturbericht und Problemskizze', in Herbert
Mehrtens and Steffen Richter (eds), *Naturwissenschaft, Technik und
NS-Ideologie. Beiträge zur Wissenschaftsgeschichte des Dritten Reichs*,
Frankfurt a. M.: Suhrkamp, 15–67.

—— (1994a), 'Irresponsible Purity: the Political and Moral Structure
of Mathematical Sciences in the National Socialist State', in
Monika Renneberg and Mark Walker (eds), *Science, Technology, and
National Socialism*, Cambridge: Cambridge University Press, 324–
38.

—— (1994b), 'Kollaborationsverhältnisse. Natur- und Technikwissen-
schaftler im NS-Staat und ihre Historie', in Christoph Meinel and
Peter Voswinckel (eds), *Medizin, Naturwissenschaft, Technik und
Nationalsozialismus. Kontinuitäten und Diskontinuitäten*, Stuttgart:
Verlag für Geschichte der Naturwissenschaften und der Technik,
13–32.

—— (1994c), 'Wissenschaftspolitik im NS-Staat – Strukturen und
regionalgeschichtliche Aspekte', in Wolfram Fischer (ed.), *Exodus
von Wissenschaften aus Berlin. Fragestellungen – Ergebnisse – Desiderate.
Entwicklung vor und nach 1933* (Forschungsbericht/Akademie der
Wissenschaften zu Berlin 7), Berlin/New York: de Gruyter, 245–66.

—— (1996), 'Mathematics and War: Germany, 1900–1945', in Paul
Forman and José M. Sánchez-Ron (eds), *National Military Establish-
ments and the Advancement of Science and Technology. Studies in 20th
Century History* (Boston Studies in the Philosophy of Science 180),
Dordrecht: Kluwer Academic Publishers, 87–134.

—— and Richter, Steffen (eds) (1980), *Naturwissenschaft, Technik und
NS-Ideologie. Beiträge zur Wissenschaftsgeschichte des Dritten Reichs*,
Frankfurt a. M.: Suhrkamp.

Meinel, Christoph and Voswinckel, Peter (eds) (1994), *Medizin,
Naturwissenschaft, Technik und Nationalsozialismus. Kontinuitäten und
Diskontinuitäten*, Stuttgart: Verlag für Geschichte der Naturwissen-
schaften und der Technik.

Merton, Robert K. (1942), 'Science and Technology in a Democratic
Order', reprinted as 'The Normative Structure of Science', in
Robert K. Merton (1973), *The Sociology of Science. Theoretical and
Empirical Investigations*, ed. Norman W. Storer, Chicago/London: The
University of Chicago Press, 267–78.

Messerschmidt, Rolf (1991), 'Nationalsozialistische Raumforschung
und Raumordnung aus der Perspektive der "Stunde Null"', in
Michael Prinz and Rainer Zitelmann (eds), *Nationalsozialismus und*

Modernisierung, Darmstadt: Wissenschaftliche Buchgemeinschaft, 117–38.

Nagel, Brigitte (1991), *Die Welteislehre. Ihre Geschichte und ihre Rolle im 'Dritten Reich'*, Stuttgart: Verlag für Geschichte der Naturwissenschaften und der Technik.

—— (1994), 'Die *Welteislehre*: Ihre Geschichte und ihre Bedeutung im "Dritten Reich"', in Christoph Meinel and Peter Voswinckel (eds), *Medizin, Naturwissenschaft, Technik und Nationalsozialismus. Kontinuitäten und Diskontinuitäten*, Stuttgart: Verlag für Geschichte der Naturwissenschaften und der Technik, 166–72.

Oexle, Otto Gerhard and Schulze, Winfried (eds) (1999), *Deutsche Historiker im Nationalsozialismus*, Frankfurt a. M.: Fischer.

Olff-Nathan, Josiane (ed.) (1993), *La science sous le Troisième Reich. Victime ou alliée du Nazisme?*, Paris: Editions du Seuil.

Oosterhuis, Harry (1997), 'Medicine, Male Bonding and Homosexuality in Nazi Germany', *Journal of Contemporary History*, 32: 187–205.

Peukert, Detlev J. K. (1986), *Grenzen der Sozialdisziplinierung. Aufstieg und Krise der deutschen Jugendfürsorge von 1878 bis 1932*, Köln: Bund-Verlag.

Prinz, Michael and Zitelmann, Rainer (eds) (1991), *Nationalsozialismus und Modernisierung*, Darmstadt: Wissenschaftliche Buchgemeinschaft; 2nd expanded edn 1994.

Proctor, Robert N. (1988), *Racial Hygiene. Medicine under the Nazis*, Cambridge, MA: Harvard University Press.

—— (1999), *The Nazi War on Cancer*, Princeton, NJ: Princeton University Press.

Rammstedt, Otthein (1986), *Deutsche Soziologie 1933–1945. Die Normalität einer Anpassung*, Frankfurt a. M.: Suhrkamp.

Raphael, Lutz (1996), 'Die Verwissenschaftlichung des Sozialen als methodische und konzeptionelle Herausforderung für eine Sozialgeschichte des 20. Jahrhunderts', *Geschichte und Gesellschaft*, 22: 165–93.

Renneberg, Monika and Walker, Mark (1994a), 'Scientists, Engineers and National Socialism', in Monika Renneberg and Mark Walker (eds), *Science, Technology, and National Socialism*, Cambridge: Cambridge University Press, 1–29.

—— (eds) (1994b), *Science, Technology, and National Socialism*, Cambridge: Cambridge University Press.

Richter, Steffen (1980), 'Die "Deutsche Physik"', in Herbert Mehrtens and Steffen Richter (eds), *Naturwissenschaft, Technik und NS-Ideologie*.

Beiträge zur Wissenschaftsgeschichte des Dritten Reichs, Frankfurt a. M.: Suhrkamp, 116–41.

Rössler, Mechtild (1990), *'Wissenschaft und Lebensraum'. Geographische Ostforschung im Nationalsozialismus. Ein Beitrag zur Disziplingeschichte der Geographie* (Hamburger Beiträge zur Wissenschaftsgeschichte 8), Berlin/Hamburg: Reimer.

—— and Schleiermacher, Sabine (eds) (1993), *Der 'Generalplan Ost'. Hauptlinien der nationalsozialistischen Planungs- und Vernichtungspolitik*, Berlin: Akademie-Verlag.

Schleiermacher, Sabine (1988), 'Die SS-Stiftung "Ahnenerbe". Menschen als Material für "exakte" Wissenschaft', in Rainer Osnowski (ed.), *Menschenversuche. Wahn und Wirklichkeit*, Köln: Kölner Volksblatt-Verlag, 70–87.

Schönwälder, Karen (1992), *Historiker und Politik. Geschichtswissenschaft im Nationalsozialismus*, Frankfurt a. M./New York: Campus.

Schöttler, Peter (1999), 'Einsatzkommando Wissenschaft. Neue Forschungen zum Verhalten deutscher Gelehrter im "Dritten Reich"', *Die ZEIT* no. 33, 12 August 1999, 39–40.

—— (ed.) (1997), *Geschichtsschreibung als Legitimationswissenschaft 1918–1945*, Frankfurt a. M.: Suhrkamp.

Siegmund-Schultze, Reinhard (1998), *Mathematiker auf der Flucht vor Hitler. Quellen und Studien zur Emigration einer Wissenschaft* (Dokumente zur Geschichte der Mathematik 10), Braunschweig/Wiesbaden: Vieweg.

Stehr, Nico (1994a), *Arbeit, Eigentum und Wissen. Zur Theorie von Wissensgesellschaften*, Frankfurt a. M.: Suhrkamp.

—— (1994b), *Knowledge Societies*, London: Sage.

—— and Ericson, Richard V. (eds) (1992), *The Culture and Power of Knowledge. Inquiries into Contemporary Societies*, Berlin: de Gruyter.

Strauss, Herbert A. et al. (eds) (1991), *Die Emigration der Wissenschaften nach 1933. Disziplingeschichtliche Studien*, München: K. G. Saur.

Szöllösi-Janze, Margit (1998), *Fritz Haber 1868–1934. Eine Biographie*, München: C. H. Beck.

—— (2000), 'Der Wissenschaftler als Experte – Kooperationsverhältnisse von Staat, Militär, Wirtschaft und Wissenschaft 1914–1933', in Doris Kaufmann (ed.), *Geschichte der Kaiser-Wilhelm-Gesellschaft im Nationalsozialismus. Bestandsaufnahme und Perspektiven der Forschung*, Göttingen: Wallstein, 46–64.

Trischler, Helmuth (1992), *Luft- und Raumfahrtforschung in Deutschland 1900–1970. Politische Geschichte einer Wissenschaft* (Studien zur

Geschichte der deutschen Großforschungseinrichtungen 4), Frankfurt a. M./New York: Campus.

—— (1994), 'Self-Mobilization or Resistance? Aeronautical Research and National Socialism', in Monika Renneberg and Mark Walker (eds), *Science, Technology, and National Socialism*, Cambridge: Cambridge University Press, 72–87.

Walker, Mark (1989a), *German National Socialism and the Quest for Nuclear Power 1939–1949*, Cambridge: Cambridge University Press.

—— (1989b), 'National Socialism and German Physics', *Journal of Contemporary History*, 24, 63–89.

—— (1995), *Nazi Science. Myth, Truth, and the German Atomic Bomb*, New York/London: Plenum Press.

Weiss, Burghard (1996), *'Forschungsstelle D'. Der Schweizer Ingenieur Walter Dällenbach (1892–1990), die AEG und die Entwicklung kernphysikalischer Großgeräte im nationalsozialistischen Deutschland*, Berlin: Verlag für Wissenschafts- und Regionalgeschichte Engel.

Wengenroth, Ulrich (1993a), 'Die Technische Hochschule nach dem Zweiten Weltkrieg. Auf dem Weg zu High-Tech und Massenbetrieb', in Ulrich Wengenroth (ed.), *Technische Universität München. Annäherungen an ihre Geschichte* (Factum 1), München: Technische Universität, 261–98.

—— (1993b), 'Zwischen Aufruhr und Diktatur. Die Technische Hochschule 1918–1945', in Ulrich Wengenroth (ed.), *Technische Universität München. Annäherungen an ihre Geschichte* (Factum 1), München: Technische Universität, 215–60.

Wolf, Ursula (1996), *Litteris et patriae. Das Janusgesicht der Historie* (Frankfurter historische Abhandlungen 37), Stuttgart: Steiner.

SYLVIA PALETSCHEK

The Invention of Humboldt and the Impact of National Socialism: The German University Idea in the First Half of the Twentieth Century

In the twentieth century, the term *Humboldtsche Universität* (Humboldtian university) became the catchword for identifying the German university system. Until the present day in German academic culture and university politics the Humboldtian ideal figures as 'creating identity *per se*'. It developed a 'force which shaped mentality' and created 'dynamics which formed reality' (Schubring/Hülten-schmidt 1991: 10). The term 'Humboldtian university' refers to the view that the modern German university emerged with the new humanistic university idea which was developed around 1800 by idealistic scholars such as Wilhelm von Humboldt, Johann Fichte and Friedrich Schleiermacher. According to the historical myth of the Humboldtian university, the founding of Berlin University in 1810 was the first manifestation of the new humanistic university idea and the later model for other German universities. It is supposed that Wilhelm von Humboldt, as Prussian minister of culture and as philosopher, played a leading role in the foundation of Berlin University (Anrich 1956; vom Bruch 1997; Boehm 1983). The concept of the Humboldtian university as it is used in today's scholarly discussion comprises the following elements: the unity of research and teaching; the function of the university as a research institution; the freedom of research and teaching which allows the university to function in furthering *pure* science (which is to say a science free of vested interests. Science in the German understanding refers to sciences and the humanities); the assumption that science provides moral education (*sittliche Menschenbildung durch Wissenschaft*); the idea that all academic disciplines should be represented within the university, and that the faculty of philosophy should function as the core, as the glue between the different disciplines and faculties.

Interestingly enough, throughout the nineteenth century the term 'Humboldtian university' was not used to characterize the German university system. In the definitions of *Universität* in encyclopaedias, books or speeches of that time, the name of Wilhelm von Humboldt does not even appear. The idea of a new humanistic university appears only marginally, if at all. From the nineteenth-century perspective, the foundation of the University of Berlin did not represent a major break with tradition. The development of the modern university was instead associated with the triumph of rationalism and organizational reform of the new *Reformuniversitäten* of Göttingen and Halle during the Enlightenment (Paletschek 2000b). According to the definitions in widely-read encyclopaedias, contemporary German universities were characterized by three responsibilities: first and above all, they were institutions of training for the academic professions; secondly they had to further develop the humanities and sciences; and thirdly they were agencies of *Allgemeinbildung* ('general education'). Freedom of teaching and research was regarded as the German university's special feature and its recipe for success. This freedom was derived as a historical tradition from the university's earlier corporate independence. It was not rooted in an idealistic philosophy.

It was only after 1900 that the origins of the modern German university began to be identified with the new humanistic university idea and the University of Berlin. But from the beginning of the twentieth century, the question of the university has been dominated by the recourse to this so-called German university idea. A long tradition of continuity from the university of the early nineteenth century to the present day has been constructed. Since then, reference to the Humboldtian ideal has served as a stratagem to justify contemporary university concepts. Between 1920 and 1960, major scholars within the German humanities such as Eduard Spranger, Carl Heinrich Becker or Helmut Schelsky postulated an ahistorical, everlasting German university idea (Spranger 1919; Spranger 1930; Becker 1925; Schelsky 1963). At the same time they played an important role in German university politics. In interpreting the writings of Humboldt, Fichte and Schleiermacher they formulated the contemporary image of the ideal university. The embedding of

Spranger's, Becker's and Schelsky's historical works in the context of concrete applications within university politics initiated the invention of the classical university, or, as I prefer to call it, the *invention of Humboldt*.

This 'invention of tradition' (Hobsbawm) underwent various metamorphoses in the first half of the twentieth century. The first stage between 1910 and 1930 embraced the enthronement and consolidation of the new humanistic university concept as a permanent ideal. The second stage under National Socialism represented the overthrow of this new humanistic university ideal. The third stage began in the period after 1945. At that time, the act of going back to Humboldt served to signify a decisive break with National Socialism. The return to the neo-humanistic university idea as a reaction to National Socialism furthered, however, the restoration of obsolete conditions at the universities after 1945.

The Invention of Humboldt – First Stage: 1910–1930

From 1910 onwards, leading scholars and university politicians showed a growing tendency to refer to the new humanistic university ideal. This can be explained first by the fact that around that time the term seems to describe the existing contemporary type of university. From the 1880s onwards, the universities had increasingly developed into research institutions (Wittrock 1993: 342).

First it was possible to legitimize this emphasis on research by referring to Humboldt's definition of research as an ongoing, incomplete project and his concept of the university as an institution of research. These essentials of Humboldt's university idea are found in his famous and unfinished essay *Über die innere und äußere Organisation der höheren wissenschaftlichen Anstalten in Berlin* ('On the internal and external organization of higher academic institutions in Berlin'). It was probably written around 1810, but unpublished und unknown until 1900. This unfinished text is 'perhaps the most discussed document in the modern history of universities'(Wittrock 1993: 317). This short essay, which has been quoted in innumerable publications on the history of universities and scholarship, was discovered in an archive by Bruno Gebhardt in the 1890s. In the nineteenth century, the contemporaries did not know this text. The whole essay was first published in 1903 (Gebhardt 1903: 250–60). All the well-known Humboldtian quotes are found in this text: *Einsamkeit und Freiheit* (loneliness and freedom) are supposed to be the driving principles of the *reine Idee der Wissenschaft* (pure idea of science). The institutions of higher scientific learning are to treat science (in the German understanding as sciences and the humanities) as 'etwas noch nicht ganz Gefundenes und nie ganz Aufzufindendes' (something not yet completely found and something which will never be recovered completely) and their task is 'unablässig sie als solche zu suchen' (to continue to search for it for ever). Extension of knowledge is a function of the universities; the progress and promotion of science in Germany is due much more to university scholars than to scholars in academies of science (Humboldt 1809–10/1982: 255, 257, 262).

Secondly, this view was supported by the leading tradition of German historiography which interpreted ideas as driving forces in history and which regarded the Prussian state as the leading force in Germany's national development. According to this view, the new humanistic university idea was brought to life with the foundation of Berlin University by the state of Prussia, which thus fulfilled its national mission and created the *true* German university. Thirdly, the turn to the new humanistic university idea was supported by the advance of neo-idealist thinking after 1900.

In the 1920s, this 'invention of Humboldt' was consolidated in the actual discussion about universities. After the First World War, demands were made for university reform in Germany since short-comings in the education system were also blamed for the defeat (Becker 1919: viii). Now problems that had already been discussed before the First World War gained a new importance. Critics complained that students were over-stressed by the sheer load of specialist knowledge and by the freedom to study what they wished. Critical contemporaries argued that the courses were too unstructured and too theoretical and that the professors spent too much time on research and paid too little attention to teaching.

A further factor was the increasing number of students in the 1920s. By 1930, there were approximately twice as many university students as in 1910. The growth of student numbers brought with it a dramatic social change. By 1930, approximately 70 per cent of all students came from the middle and lower middle classes. Female students made up about 20 per cent of the total student population. At the end of the 1920s, there was a discussion about an academic overload which presented the horror scenario of a generation of unemployed academics (Titze 1995; Titze 1990). From 1930 onwards, the universities were increasingly analysed in terms of crisis. The key terms used here were depersonalization, feminization, and the fear of an intellectual decline, which merged into a gloomy prognosis for the future.

For the majority of professors, Germany's defeat in the First World War represented a decisive turning point. Like other members of the educated upper middle classes, they mourned the loss of the constitutional monarchy. Thus, the majority of professors took on a hostile or at least a sceptical attitude towards the Weimar Republic. They saw themselves as pillars of national culture, towering over political parties and interest groups (Ringer 1969; Jansen 1992; Vogel 1991). However, there was also a smaller, but not insignificant, group of professors – frequently natural scientists – who regarded themselves as scientific experts and worked in a pragmatic manner on specific problems (Harwood 1993: 189).

This was the political and social background to the emerging discussion of university reform in the 1920s. Hardly any of the reforms proposed in the 1920s were implemented. In the discussion about the university, two positions stood out: one side wished to retain the existing university structure, the other to change universities radically. Max Scheler, for example, was among the radical critics. He held the view that the various functions of the university – professional training, research, general education – could no longer be unified within a single organization. According to Scheler, this 'functional universality' had its origins in the Middle Ages (Scheler 1926: 493, 496–502). In his view the new humanistic university idea had merely added general education (*Menschenbildung*) to the responsibilities of the university. According to Scheler, this concept had been developed at a time when neither modern research nor specialized academic education were known. Scheler argued in favour of a functional separation and the restructuring of universities into institutions for professional academic training.

Opposing this position, the second camp wished to retain the university's multiplicity of tasks. The Prussian Minister of Culture and former Professor of Oriental Studies, Carl Heinrich Becker, and the Professor of Educational Science, Eduard Spranger, were leading advocates of this position. They popularized the idea of a German university outside or above temporal or historical restrictions. Their recourse to the new humanistic university idea served as a defence of the university system's status quo against those advocating radical changes. However, they also expected that the new interpretation of the Humboldtian ideal would provide the solution of the universities' current problems. Carl Heinrich Becker wrote in 1925:

Vom Wesen der deutschen Universität kann man nur mit ehrfürchtiger
Scheu sprechen . . . Wenn wir von Universität sprechen . . . steht klar und
deutlich ein Idealbild vor der Seele, eine Art von Gralsburg der reinen
Wissenschaft. Ihre Ritter vollziehen einen heiligen Dienst.

(One can only speak of the nature of the German university with
reverential awe . . . When we speak about the university . . . we have a
clear and distinct ideal image in our souls, a sort of Holy Grail of pure
scholarship. Its knights serve a sacred cause.) (Becker 1925: 1)

In Becker's view, the German university had nothing to do with
utilitarian considerations or professional training. It pursued a
'selfless search without goals' ('selbstloses und zweckloses Suchen')
(Becker 1925: 2). Becker pleaded for the restoration of the new
humanistic university idea and a holistic approach. In his view, the
current problems of specialization, egoism and materialism were
rooted in positivism, rationalist thinking, and the era of natural
sciences and technology, which had supplanted idealism in Germany
since the 1830s. According to Becker, science had lost 'contact with
life'. This was why the 'youth of all ages' would demand an extended
concept of scientific thinking (*erweiterter Wissenschaftsbegriff*) which took
into account the desire for cohesion and included non-rationalist
impulses (Becker 1925: 22, 25). In terms of concrete measures for
reform, Becker argued for the establishment of new disciplines which
would promote intellectual synthesis, such as sociology (Müller 1991:
335–95). He argued for improved university teaching. He also
demanded more participatory rights for lecturers (*Nichtordinarien*) and
students, so that new ideas might be implemented more quickly.

The educational scientist and philosopher Eduard Spranger argued
along similar lines, but with different proposals for reform. He also
wished to return to the core of idealist thinking. He thought all
individual items of knowledge should be assembled into an 'organic
totality'. According to Spranger, the German universities had been
disrupted by the 'accelerated speed of the industrial and technological
era' (Spranger 1930: 33). In his opinion, the university crisis could
also be attributed to the democratization of education and the
'inevitable reduction in quality associated with it' (Spranger 1930:
36). The solutions proposed by him were the restriction of student
numbers and the return to the concept of the classical university on
a new basis. In response to the radical critics, he insisted that it was
still possible to link scientific research and professional training as
well as scientific research and education, and that this was the key to

success. If scientific thinking reverted from positivism to idealism, it would, in his view, also satisfy the need to provide education. In his opinion, a theoretical academic education provided better preparation for a career than professional training (Spranger 1930: 13). He felt that research without defined goals was the most useful form, as it provided solutions for unforeseen requirements which would be overlooked by scholarship based on practical matters and utilitarianism.

Basing their arguments on the neo-humanistic ideal of the university, Spranger and Becker defended the institution's traditional purposes and endorsed an elitist self-confidence among professors. In doing so, they confirmed the self-image of a university which pictured itself primarily as a research institution and which assumed that by disseminating research methodology it would simultaneously fulfil its other tasks. What they ignored was the increasing establishment of research institutions in the natural sciences outside the university since the turn of the century, which can be taken as an indication of the limitations of knowledge production within the university (Vierhaus/vom Brocke 1990). Where they took note of this development, they criticized it as a deviation from Humboldt's ideal. All in all, it seems fair to say that the recourse to an ideal picture of an eternal university model served the purpose of assigning a higher value to the humanities. Becker's and Spranger's concept of the university was drawn up entirely from the viewpoint of the humanities. The recourse to Humboldtian ideals also legitimized reforms: the deficits attributed to positivism and specialization were to be made good by a new synthetic concept.

In their reflection on universities, Becker and Spranger reacted to the yearning for a holistic world view capable of embracing rational and irrational impulses. This spirit of the time (*Zeitgeist*) was strong among students, but was also to be found in other circles of society. It characterized not only *völkisch* (nationalistic-ethnic) circles, but also reached far into the liberal, Weimar-supporting camp. It is possible to interpret the longing for a holistic world view as a reaction to rapid social change and as a response to the costs of modernization. For the traditional academic elite this modernization crisis was coupled with a loss of prestige and power, and the defeat in the First World War aggravated this crisis. There was an increasing alienation from the ideals of positivism, pluralism and parliamentarism, a strong rejection of a Western way of thinking and Western form of politics. This presented many anchorage points for the growth of National

Socialism, which promised to fulfil the widespread longings for a
holistic world view, for a cohesive national community, and for a strong
state.

The University Concept under National Socialism:
The Defeat of the Humboldtian Ideal

With the National Socialist accession to power in 1933,
the universities were integrated into the new state in a mixture of
enforced conformity and self-induced *Gleichschaltung* ('coordination').
Only very few university members offered resistance. As institutions,
the universities did not resist (Langewiesche 1997: 618; Seier 1984:
143). In 1933 the self-administration of the university was replaced
by the *Führerprinzip* (the leadership principle). Jewish staff members
and those who were members of left-wing parties were dismissed. In
total about 1,100 to 1,500 persons, that is approximately 15 per cent
of the professors and lecturers at German universities, were affected
by these measures (Ash 1995: 6). At some universities, up to 30 per
cent of the university lecturers departed without any significant
protests from their colleagues.

There are no exact figures about the university teachers who were
dismissed on political grounds by the Nazis. Mostly it is estimated
that by 1938 one third of all university teachers had been replaced
(Titze 1989: 225). But these estimates also include professors who
retired on grounds of age. Therefore these figures are too high. More
recent publications estimate that about 20 (Langewiesche 1992: 345)
or 15 per cent of the lecturers were dismissed (Fischer 1991; Ash 1995:
6). There were, however, large differences between the German
universities. While, for instance, around 30 per cent of the lecturers
at the universities of Berlin and Frankfurt were removed from office,
the dismissals at the universities of Göttingen, Hamburg and Cologne
amounted to 18–20 per cent. In comparison, only 2 per cent of the
lecturers at the university of Tübingen were dismissed (Adam 1977:
37). The differences were due to the higher or lower proportion of
Jewish scholars at these universities.

National Socialists also planned a reduction in student numbers.
They could expect that these measures would be met with approval
considering the academic overproduction that had become a problem
since 1930 (Titze 1990). In the first year after having seized power,
the National Socialists intervened in cases of new appointments to

professorships and sought to appoint professors who were convinced National Socialists. But during the following years the universities increasingly regained their ability to impose their academic standards in cases of new appointments (Adam 1977: 210–11; Seier 1983: 264). The various attempts to educate the next generation of scholars as National Socialist professors by means of training camps for lecturers (*Dozentenlager*) failed (Losemann 1980: 107; Kelly 1980: 61). From 1935 onwards, the consolidation and adaptation of these measures took place and a daily working routine in teaching and research was re-established. The outbreak of war in 1939 opened up new opportunities to some extent and also some new financial resources.

What was the stance adopted by the remaining university professors towards National Socialism? The majority of professors hoped that the Third Reich would bring about the rebirth of the nation. With this slogan, the National Socialists impressed the 'professorial seekers of the past' (*professoraler Vergangenheitssucher*; Langewiesche 1992: 376, 375–81). During the Weimar Republic these professors mourned the end of the *Kaiserreich* and the ideal of a state superior to political parties. However, the National Socialists also affected the 'anti-republican innovators' (*antirepublikanischer Erneuerer*, Langewiesche 1992: 378), who primarily comprised young scholars who criticized the fossilization of the universities. By a mixture of collaboration and private reservations the professors helped the regime to maintain itself. Yet although quite a few professors allied themselves with the new regime, the National Socialists were nevertheless sceptical of the club of professors. In their view they were remote from real life and were the representatives of a despised liberal science (Kleinberger 1980: 10). The party's ill will towards them reinforced the professors' elitist conceptions of themselves. They withdrew into ideals such as objectivity, the autonomy of the sciences and their supposedly timeless national and cultural importance. (Seier 1984: 161)

Only relatively few studies have been published on National Socialist university policies and an archive-based study is still lacking (Chroust 1993: 64; Seier 1984: 145; Heiber 1991; Losemann 1994). So far, the support of academics for National Socialism has not been studied in sufficient detail. In some cases, this support led to scientific research and teaching based on racism and *völkisch* thinking. In other cases, academic standards were maintained, but research and teaching were directed towards the aims of National Socialism (Langewiesche 1997: 620). According to the existing research, the National Socialists did not undertake a systematic reconstruction of the university system.

This is not to say that they did not intervene heavily in academic life. Their most persistent and farthest-reaching intervention was the dismissal of Jewish and leftist university lecturers and professors. Yet despite their interventions in the self-administration of the universities, they failed to introduce a new order. This lack of a fundamental university reform was caused by the brevity of the pre-war period and the competing authorities responsible for the university system (Seier 1984: 143, 148). There were no clearly defined responsibilities for the organization of the university system, although there was much wrangling about them. Many authorities were involved: the Reichserziehungsministerium (the Reich Ministry of Education), the Nationalsozialistischer Dozentenbund (the National Socialist Lecturers' Association), Himmler's Ahnenerbe (Ancestral Heritage), the university commission of the Führer's representative Rudolf Heß, the party's *Weltanschauungsbeauftragter* (official for questions of ideology) Alfred Rosenberg, as well as regional and local rulers such as the *Gauleiter* (the heads of the Nazi party's administrative districts), and university rectors. The failure of National Socialist university reform was also due to the lack of a conceptual framework (Hammerstein 1999: 118–20). National Socialist ideology offered hardly any reference points for dealing with the university. Only racial theories and *völkisch* thinking could be adopted as postulates. Instead, it is possible to detect the beginnings of a systematic National Socialist science policy in research institutes outside, rather than within, the universities – for example in the Reichsinstitut für Geschichte des neuen Deutschland (Reich Institute for the History of the New Germany) or in Himmler's Research Community, Ahnenerbe.

The political 'coordination' of the universities (Hammerstein 1999: 235–47) and the lack of fundamental reform were juxtaposed. Nevertheless, in the Third Reich a debate about university reform went on for several years without any practical outcome (Seier 1984: 148–50). It reflected divergent views. Moves towards a genuine National Socialist university concept can be detected in the project of the Hohe Schule (Bollmus 1980; Losemann 1994). This project of an alternative university was pursued from 1937 by Alfred Rosenberg, the NSDAP's official for *Weltanschauungsfragen* (questions of ideology). His alternative university was meant to carry out research and to educate future leaders for the Party and the state. However, Rosenberg did not have a mature concept of this alternative university nor did he have a strong position within the Party. He was unable to resolve the confusion over who should be responsible for university policy.

The Hohe Schule was planned for the period after the war, and was not supposed to replace the universities, but rather to take up a position above them as the 'highest institute for National Socialist research, teaching and education'. According to Rosenberg in 1940, it should provide universities 'with concretely formulated tasks ... in the sense of our *Weltanschauung*' (Bollmus 1980: 125). The research produced by the Hohe Schule was intended to link National Socialist *Weltanschauung* and science. According to Rosenberg, the experience gained in the Hohe Schule would allow for the later planning of a National Socialist university reform. From 1940 onward a few external institutes of the Hohe Schule were founded in cooperation with existing universities (Hammerstein 1999: 319), like the Institute for *Indogermanische Geistesgeschichte* in Munich, but these institutes had not much influence on further university and research policies.

Two other concepts of the university were more widely discussed than Rosenberg's model of an alternative university. They had been formulated before 1933 and were then taken up by the National Socialists in the absence of any ideas of their own. Both dealt with the problems of universities which had been discussed in the Weimar Republic. The concept of the 'political university' was propagated by Adolf Rein. Rein was a national-conservative professor of history at the university in Hamburg. His idea of the political university was supported by a large proportion of the professors and students, who saw in it a means of reinforcing the vanishing prestige of the university. The other, more radical concept originated with Ernst Krieck, who had been an influential educator in the 1920s and 1930s. Krieck, who might best be described as an anti-democratic innovator, demanded an anti-bourgeois and anti-positivist university. Both professors forged a career through National Socialism, received university professorships and became university rectors. They sought to implement the plans for reform at their universities in Hamburg and Heidelberg, but these efforts petered out. Inasmuch as the National Socialists took up both these concepts, they were pursuing different types of reform. Thus both the conservative 'seeker of the past' and the 'anti-republican innovator' found support.

A common element in the university concepts of Krieck and Rein is that they sought to overcome the new humanistic university ideal but in fact adopted parts of it. They pleaded for a final commitment to scientific research and university teaching, against the ideals of absolute freedom to choose the objects of research (*Zweckfreiheit*). At the same time, however, they wished to permit academic freedom within this framework. They spoke out against research and teaching steered by direct practical or state interests. Both maintained the idealistic concept of education through research. Rein and Krieck, like Spranger and Becker, traced the faulty development of the university to the defeat of idealism by positivism and the natural sciences around the 1830s.

With his concept of a political university Adolf Rein disassociated himself from the humanistic university of the nineteenth century and the mass university with its tendency towards the 'Americanized adult education classes'. The national-convervative Rein complained that the unity of scholarship had been lost and that 'department stores of specialized science' had been created (Rein 1933: 20). Following Max Weber, he argued that the sciences and the humanities cannot give answers to fundamental questions of life, but merely provide methods of making rational decisions. For Rein it was therefore necessary for every scholar to choose his ultimate set of values, whether based on religion, a philosophical system or politics. In his view, the guiding principle of the age could only be that of politics. Thus, the political university was the university which was aligned with the state. Nevertheless, this university should not only serve the purpose of professional training and be slavishly devoted to the interests of the state. It should work in a critical manner while still being basically committed to the state. The centre of the proposed political university was supposed to be the new faculty of politics, where humanities and social sciences were to be combined. Rein was a convinced opponent of democracy. He was committed to a national-conservative view, which was widespread among professors in the 1920s (Vogel 1991: 43), but radicalized it into an alignment of the university with the state and the nation. He considered the idea that the university might hold a neutral world view to be a fiction of the liberal and natural scientific era. He also felt that in historical reality the new humanistic university had never been dedicated solely to reason, but rather that it had been – despite all denials – tied to its religious, political and sociological context. More precisely, that it had been tied to Protestantism, the liberal-constitutional nation-state and the bourgeoisie (Rein 1933: 6). Rein's concept of the political university was not realized, as there were no corresponding initiatives from the universities themselves. He himself tried to implement his concept while he was rector of the university at Hamburg, but failed (Giles 1980: 57).

In contrast with Rein, Ernst Krieck criticized bourgeois educational privileges and the existing elitist educational ideology. Krieck came from a lower middle-class background; he was an elementary school-teacher and largely self-educated. In the 1920s and 1930s he became an influential pedagogue (Müller 1978). Disappointed because his demands with regard to educational policies had not been met in the 1920s, Krieck turned to national-revolutionary circles and joined the NSDAP as early as 1932. In 1933 he gained a chair, and was rector of the universities of Frankfurt and Heidelberg, where he tried, with relatively little success, to make progress with university and academic reform. In 1938 he had a fierce dispute with race theorists. Thereupon he resigned from his offices in the Party and at the universities and retired as an embittered man.

Krieck criticized the upper class's educational privileges, which, according to him, were secured by a hierarchical educational system and an elitist theory of education (Müller 1978: 440). In his view, the educated elites had lost touch with the present and the needs of society. He hoped that National Socialism would establish a realm of human dignity and humanity which would guarantee the right to 'endless education' and would thus redeem the promises of the Enlightenment and liberalism. He felt that contemporary universities had no contact with real life. In Krieck's opinion, science was 'of use to nobody and nothing, without meaning, without educational purpose, without ethos, surviving on the basis of a traditionally fostered prejudice' (Krieck 1932: 162). He felt that science had lost its synthetic power with the downfall of idealism. It no longer had any educational value and restricted itself to technical utilization. Only via an external link to a *völkisch-politische Weltanschauung* would science once again regain its educational function. This would put an end to 'the liberalistic illusion, arbitrariness, the senseless and disconnected plurality' (Krieck 1932: 164).

Unlike Rein, Krieck pleaded for the dissolution of the existing universities into seperate institutions for professional training, each with its own special branches. Each special branch of study should follow its particular perspective, always keeping in mind the linking concept, the nation, so that a connection between all disciplines could be achieved. The university professors should once again concentrate more on teaching, and not merely see it as an appendix to research. The universities should primarily provide professional training, whereas research should be done at the academies. In 1935, after it was seen that university reform 'from above' was out of the question, Krieck, having become rector of the university of Heidelberg, tried to experiment with his idea of university reform. Various study groups thus tried to achieve 'a critical distance from the spiritual tradition of the German university linked with the name of Humboldt and their own, mostly upper-class past'. At the university of Heidelberg a 'cultural offensive' against 'western ideology' should be started (Müller 1978: 125). Krieck criticized the philistinism of the technologists, the cheapening of the humanities in the Third Reich, as well as the preferential treatment of the natural and technological disciplines within the framework of the four-year plan of 1936. The educational function of science should not be permitted to be stifled by the technical function (Müller 1978: 131).

The concepts of Rein and Krieck grew out of the discussion about universities of the 1920s. They were intended as a renunciation of the new humanistic university, but ultimately retained its concept of education through science, which they wished to achieve in a new way. They also held on to the freedom of research and teaching but within a framework in which the universities were explicitly linked to the goals of the state and the nation. This could be used by the National Socialists as an instrument for the accomplishment of their political ideas, although these concepts were not realized in the end. The absence of a thoroughly formulated National Socialist university concept and the vague guidelines of racist and *völkisch* thinking provided considerable scope for interpretation by convinced National Socialists and those professors who wished to conform. With great inventiveness they sought specific solutions for linking National Socialist *Weltanschauung* and science. On the basis of the *Führerprinzip* ('leader principle'), those who were rectors had the power to experiment in implementing new university concepts. Thus, the rector of Tübingen University, for example, the convinced National Socialist and psychiatrist, Hermann F. Hoffmann, attempted to produce a synthesis of *Weltanschauung* and science by means of his 'biological *Weltbild*' (Leonhardt 1996: 61; Langewiesche 1997: 644–6). He considered this *Weltbild*, which was rooted in racism, to be a guideline for the humanities. He altered the traditional precedence of the faculties. The faculty of theology no longer occupied the first place, but instead, he transferred the faculty of natural sciences, which was last in the traditional hierarchy, to the top (Hoffmann 1940: 114–26). Thus, the faculty of natural sciences was placed at the centre of his university. It remains unclear to what extent Hoffmann's view of the university was typical for natural or medical scientists. It appears that in public the National Socialists did not propagate a concept of the university based on the perspective of the natural sciences, although they invested much money in research projects on natural sciences outside the universities.

The debate about universities and the development of the universities under National Socialism can be briefly summarized as follows: The severe losses of German university teachers after 1933 due to dismissals, persecution and emigration meant in the long run a loss of innovation and cultural diversity in the German university scene. A modernization of the universities did not take place. It was only in the first years after the Nazi takeover that some small steps into this direction were discernible. However, these attempts either fell flat or were abandoned, being seen as suspicious democratic tendencies within the university. These modern elements consisted of the initially planned and partially implemented participation of students in the universities' self-governing bodies, the extended opportunities for lecturers below the ranks of full-professors to exert influence, and the creation of material security for them. All this, however, was only achieved at the expense of conformity with the system. The main reason for the National Socialists' attempts to break the power of the established professors in the universities was their expectation that they could achieve a National Socialist orientation of the universities if more influence were given to the younger elements, who were more likely to hold National Socialist views than the older professors. Some younger professors quickly obtained chairs and influential positions at universities through their political offices, for example as *Dozentenführer* ('lecturers' leaders'). However, in the daily routines of the universities the institution's inertia soon began to show again. The introduction of the 'leader principle' suspended the traditional collegial principle of the university's self-administration, but the rule of the full professors was maintained. There were even tendencies towards its consolidation.

When the National Socialists seized power they did not have a clearly defined concept of university reform. Their vague guidelines could be interpreted rather broadly and therefore seem to have encouraged a range of different expectations of the professoriate. The debate on the university under National Socialism must not be perceived as completely distinct from that of the first half of the twentieth century, since the National Socialists took up concepts which had developed from the discussions of the 1920s. And the ideas about university reform after 1945 were, ultimately, formulated in order to disassociate universities from their development under National Socialism. On quite different premises and imbued with quite different opinions than in the 1920s, this again led to an idealization of the neo-humanistic university idea. This ideological papering-over of university problems helped to ossify conditions at German universities until the 1960s and beyond.

The 'Invention' of Humboldt - Second Stage: 1945–1960

After the end of the Second World War and the collapse of the National Socialist regime there was a return to the Humboldtian ideal in the discussion about universities and in the legitimization of university reform – in both East and West Germany. In the Soviet Occupied Zone and the later GDR, university policies were presented as bridging the gap back to the Humboldt tradition (Connelly 1997). A symbolic expression of this continuity was the renaming of the Friedrich-Wilhelms-Universität Berlin as the Humboldt-Universität in rememberance of Alexander and Wilhelm von Humboldt and 'in recognition of what has so far been achieved in the fight for the democratic university and as a commitment for the future' (Humboldt Universität 1960: 132). Now, a democratic inheritance was derived from the neo-humanistic tradition. The recourse to Humboldt legitimized the political guidelines of Soviet university policies, the elimination of fascism and militarism, and the transmission of basic democratic principles as well as the opening up of bourgeois educational privilege. However, it failed to recognize the break with the tradition of the research university, which was resulting from a tendency towards a gradual transfer of research away from universities to the academies of science.

The return to the 'everlasting' (*überzeitlich*) university ideal of Humboldt was taken far more seriously in West Germany and has left its mark on the discussion about universities up to the present day (Jarausch 1997 and 1999). The new humanism and classics were understood as a form of safeguard against National Socialism, which had broken with this cultural tradition. After 1945, issues discussed in the 1920s reappeared, showing the restaurative tendencies in post-war Germany. The well-known *Blaues Gutachten* (Blue Audit) of 1948 demanded above all a renewal of the educational functions of the universities (Gutachten 1948). An attempt was made to introduce a *studium generale* and to lead the students in the direction of a humanistic education, so that the barbarism of National Socialism would never be repeated. This proposed *studium generale* was not really accepted by the students und was withdrawn in the mid-1950s (Studium Generale 1951).

The reference to the Humboldtian ideal allowed a multiplicity of university responsibilities. It reinforced the self-image of the university as a research institution and made it possible to link it up with supposedly democratic traditions in German history. The reaction to National Socialism led to a renewed fixation with the university ideal of neo-humanism, which had been propagated as timeless since the 1920s. Linked with this right up to the 1960s was a desire to preserve university institutions which had already existed in the 1920s. The National Socialist rejection of the timeless concept of the German university discredited from 1945 onwards any proposed deviation from the Humboldtian ideal. National Socialism was presented as a sharp break in the history of the German university, as a time of ruin into which the university was forced for twelve years, as Karl Jaspers put it in 1945 (Jaspers 1945/1986: 103).

From 1945 onwards a denazification of the universities took place, but with the exception of a small number of confirmed National Socialists the majority of professors was soon lecturing and carrying out research again. This corresponded with the common formulation which had been conjured up in many speeches and which had already been used in the 1920s after the defeat in the First World War: 'The core of our universities is healthy' (*Der Kern unserer Universitäten ist gesund*) (Becker 1919: 17; Gutachten 1948: 291). According to Jaspers 'it has not yet been possible to destroy the academic spirit' (Jaspers 1945/1986: 103). Until the 1960s and in some aspects until the 1980s, this idea prevented further questions about the previous academic alignment with National Socialism and with racist and *völkisch* premises. Though many professors had conformed to National Socialism during the Third Reich, they could now hide behind the 'timeless' ideal of Humboldt.

The reaction to National Socialism signified one more stage in the dissemination of the Humboldtian university ideal in the discourse about universities. Further stages of the invention of Humboldt followed. At the beginning of the 1960s for instance, a third stage became apparent. Here, Helmut Schelsky's book *Einsamkeit und Freiheit* (Loneliness and Freedom) played a pivotal role (1963). Schelsky took up a postulate of the sociologist René König, which was adduced by the latter in defence against the political university of National Socialism. 'The idea of the university as it was put forward in German

idealism is the normative framework by which every true German
university reform will have to be verified' (König 1935: 13). For
Schelsky, this meant that each renewal of the university in Germany
had to take place within the normative framework of the neo-
humanistic idea. As late as 1963 Schelsky described the foundation
of Berlin University 1810 as the 'present past' (*aktuelle Vergangenheit*,
Schelsky 1963: 48). In his demands for a contemporary university
reform Schelsky referred to the ideas of Wilhelm von Humboldt,
reinterpreting them and emphasizing the research imperative. He
called for a concept of scientific education in keeping with the times
and a new theory of science as starting point of a university reform
and the intended *theoretical* university. He rejected, however, a view of
education and science derived from the Humboldtian ideas after 1945
which was 'merely idealistic, only rooted in the history of ideas' ('eine
nur ideenhaft-geistesgeschichtliche Vorstellung von Bildung und
Wissenschaft', Schelsky 1963: 8)

Argumentation in the supposed tradition of Humboldt extends well
into the present-day university debate. It is used by defenders of the
existing university system as well as by its critics and reformers. Here,
a long arch of continuity from the neo-humanistic university idea and
the foundation of Berlin University in 1810 to the present is drawn.
This arch of continuity covers the manifold changes the university
has experienced in the nineteenth and twentieth centuries. As a result,
in the twentieth century a discussion of the university's self-concept
was only possible by an intricate recourse to the presumed positions
of the early nineteenth century.

References

Adam, Uwe Dietrich (1977), *Hochschule und Nationalsozialismus. Die
    Universität Tübingen im Dritten Reich*, Tübingen: Mohr.
Anrich, Ernst (1956), *Die Idee der deutschen Universität. Die fünf Grund-
    schriften aus der Zeit ihrer Neubegründung durch klassischen Humanismus
    und romantischen Idealismus*, Darmstadt: Wissenschaftliche Buch-
    gesellschaft.
Ash, Mitchell G. (1995), 'Wissenschaftswandel in Zeiten politischer
    Umwälzungen: Entwicklungen, Verwicklungen, Abwicklungen',
    *Internationale Zeitschrift für Geschichte und Ethik der Naturwissenschaften,
    Technik und Medizin*, 3: 1–21.
Becker, Carl Heinrich (1919), *Gedanken zur Hochschulreform*, Leipzig:
    Quelle & Meyer.

—— (1925), 'Vom Wesen der deutschen Universität', in Reinhold Schairer and Conrad Hofmann (eds), *Die Universitätsideale der Kulturvölker*, Leipzig: Quelle & Meyer, 1–30.

Boehm, Laetitia (1983), 'Wilhelm von Humboldt (1767–1835) and the University: Idea and Implementation', *CRE-Informationen*, 62: 89–105.

Bollmus, Reinhard (1980), 'Zum Projekt einer nationalsozialistischen Alternativ-Universität: Alfred Rosenbergs "Hohe Schule"', in Manfred Heinemann (ed.), *Erziehung und Schulung im Dritten Reich. Teil 2: Hochschule, Erwachsenenbildung*, Stuttgart: Klett-Cotta, 125–52.

Chroust, Peter (1993), 'Deutsche Universitäten und Nationalsozialismus', in Jürgen Schriewer et al. (eds), *Sozialer Raum und akademische Kulturen*, Frankfurt: Lang, 61–112.

Connelly, John (1997), 'Humboldt Coopted: East German Universities, 1945–1989', in Michell G. Ash (ed.), *German Universities Past and Future. Crisis or Renewal?* Providence: Berghahn, 55–76.

Fischer, Klaus (1991), 'Die Emigration von Wissenschaftlern nach 1933: Möglichkeiten und Grenzen einer Bilanzierung', *Vierteljahrshefte für Zeitgeschichte*, 39: 535–49.

Gebhardt, Bruno (ed.) (1903), *Wilhelm von Humboldts politische Denkschriften*, vol. 1: *1802–1810*, Berlin: Behr.

Giles, Geoffrey J. (1980), 'Die Idee der politischen Universität. Hochschulreform nach der Machtergreifung', in Manfred Heinemann (ed.), *Erziehung und Schulung im Dritten Reich. Teil 2: Hochschule, Erwachsenenbildung*, Stuttgart: Klett-Cotta, 50–60.

Gutachten zur Hochschulreform vom Studienausschuß für Hochschulreform ('Blaues Gutachten') (1948), in Rolf Neuhaus (ed.) (1961), *Dokumente zur Hochschulreform 1945–1959*, Wiesbaden: Steiner, 289–368.

Hammerstein, Notker (1999), *Die Deutsche Forschungsgemeinschaft in der Weimarer Republik und im Dritten Reich. Wissenschaftspolitik in Republik und Diktatur*, München: C. H. Beck.

Harwood, Jonathan (1993), '"Mandarine" oder Außenseiter? Selbstverständnis deutscher Naturwissenschaftler (1900–1933)', in Jürgen Schriewer et al. (eds), *Sozialer Raum und akademische Kulturen. Studien zur europäischen Hochschul- und Wissenschaftsgeschichte im 19. und 20. Jahrhundert*, Frankfurt: Lang, 183–212.

Heiber, Helmut (1991), *Universität unterm Hakenkreuz, Teil I: Der Professor im Dritten Reich. Bilder aus der akademischen Provinz*, München: K. G. Saur.

—— (1992/1994), *Universität unterm Hakenkreuz, Teil II,1/2: Machter-greifung und Gleichschaltung im akademischen Raum*, München: K. G. Saur.

Hoffmann, Hermann F. (ed.) (1940), *Universität Tübingen 1938–1939*, Tübingen.

Humboldt, Wilhelm von (1809–10/1982), 'Über die innere und äußere Organisation der höheren wissenschaftlichen Anstalten in Berlin (1809/10)', in Andreas Flitner and Klaus Giel (eds), *Wilhelm von Humboldt. Schriften zur Politik und zum Bildungswesen*, vol. IV, Darm-stadt: Wissenschaftliche Buchgesellschaft, 255–66.

Humboldt-Universität (1960), *Die Humboldt-Universität gestern, heute, morgen. Zum einhundertfünfzigjährigen Bestehen der Humboldt-Universität zu Berlin und zum zweihundertfünfzigjährigen Bestehen der Charité*, Berlin: Deutscher Verlag der Wissenschaften.

Jansen, Christian (1992), *Professoren und Politik. Politisches Denken und Handeln der Heidelberger Hochschullehrer 1914–1935*, Göttingen: Van-denhoeck & Ruprecht.

Jarausch, Konrad H. (1997), 'The Humboldt Syndrom: West German Universities, 1945–1989 – An Academic *Sonderweg?*' in Michell G. Ash (ed.), *German Universities Past and Future. Crisis or Renewal*, Providence: Berghahn, 33–49.

—— (1999), 'Gebrochene Traditionen: Wandlungen des Selbstver-ständnisses der Berliner Universität', *Jahrbuch für Universitätsgeschichte*, 2: 121–36.

Jaspers, Karl (1945/1986), 'Erneuerung der Universität (Rekto-ratsrede 1945)', in Karl Jaspers (1986), *Erneuerung der Universität. Reden und Schriften 1945/46*, Heidelberg: Schneider, 93–106.

—— (1946), *Die Idee der Universität*, Berlin: Springer.

Kelly, Reece C. (1980), 'Die gescheiterte nationalsozialistische Personalpolitik und die mißlungene Entwicklung der national-sozialistischen Hochschulen', in Manfred Heinemann (ed.), *Erziehung und Schulung im Dritten Reich. Teil 2: Hochschule, Erwachsenenbildung*, Stuttgart: Klett-Cotta, 61–76.

Kleinberger, Aharon F. (1980), 'Gab es eine nationalsozialistische Hochschulpolitik?', in Manfred Heinemann (ed.), *Erziehung und Schulung im Dritten Reich. Teil 2: Hochschule, Erwachsenenbildung*, Stuttgart: Klett-Cotta, 9–30.

König, René (1935), *Vom Wesen der deutschen Universität*, Berlin: Die Runde.

Krieck, Ernst (1932), *Nationalpolitische Erziehung*, Leipzig: Armanen Verlag.

Langewiesche, Dieter (1992), 'Die Eberhard-Karls-Universität Tübingen in der Weimarer Republik. Krisenerfahrungen und Distanz zur Demokratie an deutschen Universitäten', *Zeitschrift für Württembergische Landesgeschichte*, 51: 345–81.

—— (1997), 'Die Universität Tübingen in der Zeit des National-sozialismus: Formen der Selbstgleichschaltung und Selbstbehauptung', *Geschichte und Gesellschaft*, 23: 618–46.

Leonhardt, Martin (1996), *Hermann F. Hoffmann (1891–1944). Die Tübinger Psychiatrie auf dem Weg in den Nationalsozialismus*, Sigmaringen: Thorbecke.

Losemann, Volker (1980), 'Zur Konzeption der NS-Dozentenlager', in Manfred Heinemann (ed.), *Erziehung und Schulung im Dritten Reich. Teil 2: Hochschule, Erwachsenenbildung*, Stuttgart: Klett-Cotta, 87–109.

—— (1994), Reformprojekte der NS-Hochschulpolitik, in Karl Strobel (ed.), *Die deutsche Universität im 20. Jahrhundert. Die Entwicklung einer Institution zwischen Tradition, Autonomie, historischen und sozialen Rahmenbedingungen*, Vierow: SH-Verlag, 97–115.

Müller, Gerhard (1978), *Ernst Krieck und die nationalsozialistische Wissenschaftsreform*, Weinheim: Beltz.

Müller, Guido (1991), *Weltpolitische Bedeutung und akademische Reform. Carl Heinrich Beckers Wissenschafts- und Hochschulpolitik 1908–1930*, Köln: Böhlau.

Paletschek, Sylvia (2000a), *Die permanente Erfindung einer Tradition. Studien zur Geschichte der Universität Tübingen im Kaiserreich und in der Weimarer Republik*, Stuttgart: Steiner.

—— (2000b), 'Verbreitete sich ein Humboldtsches Modell an den deutschen Universitäten im 19. Jahrhundert?', in Rainer Christoph Schwinges (ed.), *Humboldt international. Der Export des deutschen Universitätsmodells im 19. und 20. Jahrhundert*, Basel: Schwabe.

Rein, Adolf (1933), *Die Idee der politischen Universität*, Hamburg: Hanseatische Verlagsanstalt.

Ringer, Fritz K. (1969), *The Decline of the German Mandarins. The German Academic Community, 1890–1933*, Cambridge: Harvard University Press.

Scheler, Max (1926), *Die Wissensformen und die Gesellschaft*, Leipzig: Der-neue-Geist-Verlag.

Schelsky, Helmut (1963), *Einsamkeit und Freiheit. Idee und Gestalt der deutschen Universität und ihrer Reformen*, Reinbek: Rowohlt.

Schubring, Gert/Hültenschmidt, Erika (1991), 'Vorwort', in Gert Schubring (ed.), *'Einsamkeit und Freiheit' neu besichtigt. Universitäts-*

reformen und Disziplinenbildung in Preußen als Modell für Wissenschaftspolitik im Europa des 19. Jahrhunderts, Stuttgart: Steiner.

Seier, Hellmut (1983), 'Die Hochschullehrerschaft im Dritten Reich', in Klaus Schwabe (ed.), *Deutsche Hochschullehrer als Elite 1815–1945*, Boppard: Harald Boldt Verlag, 247–95.

―― (1984), 'Universität und Hochschulpolitik im nationalsozialistischen Staat', in Klaus Malettke (ed.), *Der Nationalsozialismus an der Macht. Aspekte nationalsozialistischer Politik und Herrschaft*, Göttingen: Vandenhoeck & Ruprecht, 143–65.

Spranger, Eduard (1919), *Über das Wesen der Universität*, Leipzig: Meiner.

―― (1930), Das Wesen der deutschen Universität', in Michael Doeberl et al. (eds), *Das akademische Deutschland*, vol. III, Berlin: Weller, 1–38.

Studium Generale (1951), in Rolf Neuhaus (ed.) (1961), *Dokumente zur Hochschulreform 1945–1959*, Wiesbaden: Steiner Verlag, 387–99.

Titze, Hartmut (1989), 'Hochschulen', in Dieter Langewiesche and Karl-Heinz Tenorth (eds), *Handbuch der deutschen Bildungsgeschichte*, vol. V: 1918–1945, München: C. H. Beck, 209–58.

―― (1990), *Der Akademikerzyklus. Historische Untersuchungen über die Wiederkehr von Überfüllung und Mangel in akademischen Karrieren*, Göttingen: Vandenhoeck & Ruprecht.

―― (ed.) (1995), *Datenhandbuch zur deutschen Bildungsgeschichte, vol. I, 2: Wachstum und Differenzierung der deutschen Universitäten 1830–1945*, Göttingen: Vandenhoeck & Ruprecht.

Vierhaus, Rudolf and vom Brocke, Bernhard (eds) (1990), *Forschung im Spannungsfeld von Politik und Gesellschaft. Geschichte und Struktur der Kaiser-Wilhelm-/Max-Planck-Gesellschaft*, Stuttgart: dva.

Vogel, Barbara (1991), 'Anpassung und Widerstand. Das Verhältnis Hamburger Hochschullehrer zum Staat 1919 bis 1945', in Eckhart Krause, Ludwig Huber and Holger Fischer (eds), *Hochschulalltag im 'Dritten Reich'. Die Hamburger Universität 1933–1945* (Hamburger Beiträge zur Wissenschaftsgeschichte 3), Berlin: Reimer, 3–84.

vom Bruch, Rüdiger (1997), 'A Slow Farewell to Humboldt? Stages in the History of German Universities, 1810–1945', in Mitchell G. Ash (ed.), *German Universities Past and Future. Crisis or Renewal?* Providence: Berghahn, 3–27.

Wittrock, Björn (1993), 'The Modern University: the Three Transformations', in Sheldon Rothblatt and Björn Wittrock (eds), *The European and American University since 1800*, Cambridge: Cambridge University Press, 303–62.

MECHTILD RÖSSLER

Geography and Area Planning under National Socialism

Over the past ten years the relationship between geography and National Socialism has received some attention. Discussion has been based on a number of research projects, among them a working group 'Geography and National Socialism', which published a book of case studies on geographical institutes (Fahlbusch/Rössler/Siegrist 1988), followed by a research project on geography as a discipline from 1933 to 1945, financed by the German Research Foundation, and finally a research project on the University of Hamburg and its history, which included geography and its applications in colonial research and area planning.

When reviewing the conclusions and the literature of the past ten years, one can note several tendencies. During the 1980s, studies of individuals and particular institutions were carried out. In the 1990s, research focused on applied geography and area research within the National Socialist 'living space' policy (including the 'General Plan East') on the one hand, and on the so-called brains-trust of *Volkstum* research and its multi-faceted organizations on the other. This was debated at the German *Historikertag* held in Frankfurt in September 1998 during a session on 'Historians and National Socialism'.[1]

Introduction

Land, sea and most recently air space constitute the ways in which the state organizes its territorial power over a populated part of the earth's surface. Geographical political discourse was a product of nineteenth century's colonial imperialism. In the course of the twentieth century this discourse assumed the form of geopolitics in the broadest sense, which in

59

turn was directly associated with the apparatus of the state, and
represented the strategic articulation of the relationship between political
power and the organization of the earth's surface. (Prigge 1986: 99, my
translation)

In relation to this geographical-political discourse, my essay will focus
on the practice of geography and area research as a discipline,
geographers as experts, and their profession's behaviour during the
creation of National Socialist 'living space' in the East. Finally, I shall
briefly review the newly evolving colonial geography in the Third
Reich and the SS expeditions carried out during the war. I shall also
discuss the situation of the discipline after the end of the Third Reich.
Specifically, I would like to review three different issues:

–   firstly, the ideological sphere, that is, the intellectual origins of
    geographical concepts, with reference to the ever wider public
    debate on German 'living space' during the Weimar Republic,
    and their place within National Socialist Ideology;
–   secondly, how these concepts were transformed and put into
    practice under the political influence of academic experts during
    the period from 1933 to 1945, that is, the realization of these
    ideas during the Second World War and the expansion to the East;
–   thirdly, the political situation immediately after the war in 1945,
    when the same geographers were hired by American agencies to
    draw the maps of divided Germany and to be involved in the
    creation of a post-war Europe.

Geographical Concepts in the Weimar Republic and the Ideological Origins of Nazi Imperialism (see Smith 1986)

From the turn of the century, geographers both developed
and helped shape concepts which after 1933 had considerable public
and political utility. All these complex developments can be described
as a form of geographical political discourse. Under the Nazi regime,
geographers shaped one of the most controversial and dubious
concepts ever to emanate from the discipline as a whole, namely
Lebensraum ('living space'), a concept which Friedrich Ratzel (1844–
1904) had introduced into the political geographical discourse. The
concept was adopted by numerous academic geographers and political

pundits and, in the form of a popular political slogan within National Socialist ideology, it assumed practical importance for expansionist policies in the East. I would like to concentrate upon the specific relationships between the key concepts and metaphors *Lebensraum*, *Drang nach Osten* ('Push to the East') and *Volks- und Kulturboden* ('people's and culture soil'),[2] the latter created by the eponymous Stiftung für Volks- und Kulturbodenforschung (Foundation for Research into the Land and Culture of the Folk).

These three concepts enjoyed wide use under the Nazi regime, and by the same token among contemporary geographers. During the Weimar Republic, geographers were much in evidence in a number of important expansionist lobby groups founded already in the Wilhelmine Empire, among them the German Colonial Society (Deutsche Kolonialgesellschaft) and the Pan-German League (Alldeutscher Verband) (see Chickering 1984), many of which were either led or influenced by prominent academic geographers. Politically, they gathered in right-wing conservative parties like the annexationist Fatherland Party (Vaterlandspartei), founded in 1917, or later the German National People's Party (Deutschnationale Volkspartei, DNVP). Their academic work increasingly corresponded to their political objectives. Public political discussion of *Lebensraum* affected academic work in this area, as the concept was gradually transformed by the new ideology. This was particularly the case with racial biological ideas, which originally had no connection with *Lebensraum*, except in so far as racism was more or less latent in all imperialist ideologies of the nineteenth century. The basic assumptions of *Lebensraum* focused on culture and environment, but the functional link with biological racism was made in the 1920s (Smith 1986: 212). In the course of its adoption as an aspect of Nazi ideology, the concept became intertwined with two other crucial elements, first the racist doctrine of a *Herrenvolk*, a superior race (the Nordic Germanic tribe over all other races), and secondly, its direction to the East. The final form of the Nazi concept of *Lebensraum* was then developed during the Third Reich with the occupation of Poland and the Soviet Union (Smith 1986; Burleigh 1988; Rössler 1990).

The Foundation for *Volks- und Kulturbodenforschung* (see for the following Fahlbusch 1994: 71–2) was created in Leipzig with the political and financial assistance of the Reich Ministry of the Interior (Reichsinnenministerium). It was based on preparatory work by bodies such as the Volksdeutsche Mittelstelle. The founding session took place in Berlin on 30 October 1923. The main purpose of this organization was 'research on German *Volks- und Kulturboden* for the best of Germandom'. It was one of the main institutions in the interplay between German nationalist politics and scientific research during the Weimar Republic. This can be illustrated by a letter from the Minister of the Interior to the Secretary-General, indicating that the work of the foundation could provide the chance 'to expose publicly the injustices of the Versailles Diktat (Dictated Peace) on national borders to the international scientific world'. A great number of geographers participated in its activities, which included biannual conferences and numerous publications. Among the publications, the journal *Hefte für deutsche Volks- und Kulturbodenforschung* and the diction-ary *Handwörterbuch für das Grenz- und Auslandsdeutschtum* should be mentioned. Although the foundation had to be closed before the Nazi takeover in 1933 owing to internal conflicts and financial problems, the activities continued within geographical institutes at universities and other organizations (Rössler 1990: 54-5).

The theoretical concept of the German *Volks- und Kulturboden* (*Boden*, 'soil') contained three different 'territories': first, the German *Reich*, in principle within the state borders; secondly, the German *Volksboden* ('ethnic territory'), which meant a wider area mainly settled by German people; and thirdly, the *Kulturboden* ('cultural area'), an even wider area, where German cultural influence in the broadest sense was predominant. The latter stood in marked contrast to the actual political frontiers. It was of fundamental importance to geographical research until 1945.

The theory of the German *Volks- und Kulturboden* became a popular concept and was widely used by scholars from other disciplines, including anthropologists, ethnographers and historians. Geographers were multipliers and catalysts during the Weimar Republic through their work in cultural-political organizations. The manager of the Leipzig foundation, Wilhelm Volz, for example, was an active member of the DNVP and the Verein für das Volkstum im Ausland (Association of German Folk Abroad). Based on their scientific groundwork and by promoting these concepts in the public sphere, these social scientists paved the way for the spatial ideologies of National Socialism.

The Nazi takeover of the German universities had no major impact on geography as a discipline. On the contrary, most geographers welcomed the new regime. Only one geographer went into exile to the United States, Emil Waibel, and another one, Alfred Philippson, was deported to the Theresienstadt concentration camp (1942–45), where he survived the Third Reich. Geography as a discipline at universities continued without any radical change or upheaval. New emphases were set in particular at the so-called 'frontier universities' with seminars on *Grenz- und Auslandsdeutschtum* (Border and Overseas Germans), for example at the University of Freiburg with a view towards Alsace (Rössler 1983).

The question of continuity and discontinuity can be discussed best by using the following example of a debate during the Second World War.

Scientific and Political Discourse: The Geographers' Debate on *Lebensraum* in 1942

At the time of the greatest extension of the German Reich in 1942, geographers started a debate on the concept of *Lebensraum* in the well-known German geographical periodical *Geographische Zeitschrift*. They believed that it was essential in these years to use their knowledge to establish the new borders of the Greater German Reich (*Großraum*), to define the exact status of the conquered territories in the East, and to reflect the various concepts of *Lebensraum* and *Großraum*. The debate was opened by Ernst Friedrich Flohr with 'An attempt to clarify the term *Lebensraum*' (1942), suggesting that this term had been transformed by recent events. The author defined a hierarchy of peoples, concluding that a living space existed only for a true *Volk*. The Jews, for example, were for him not to be regarded as a *Volk* because 'they did not belong to a particular place'. In Flohr's view, as every 'healthy' *Volk* wanted to grow, its *Lebensraum* would have to expand as well. He warned against confusing 'true' and 'false' growth. The latter, he considered an economically defined 'complementary space' (*Ergänzungsraum*) which delivered food and other goods to the motherland, but was ultimately alien to its inhabitants. African colonies were for him a typical example of an *Ergänzungsraum*. Flohr also denied that any kind of 'apparent living space' (*Scheinlebensraum*), created by foreign trade and economic exchange, would mean a healthy and stable development.

Entering the subject from the perspective of earlier academic debates, the renowned geographer Heinrich Schmitthenner put forward his own concept of *Lebensraum* (1938; 1942). He distinguished an 'active' and a 'passive' living space, and a *Lebensspielraum*, an even wider area in which the manifold 'life' of a people with all its economic, social, political and cultural relations took place. He argued for a new

Occidental *Großraum*, a major region tying together Europe and tropical Africa, which was said to be Europe's natural *Ergänzungsraum* (see also Sandner/Rössler 1994). Erich Obst, another geographer, also approved of the vision of a major regional block which would include Europe and Africa (1941), but Flohr refused to use the term *Lebensraum* for such a large area. He argued that a *Großraum* meant a *Lebensraum* community of different peoples (*Lebensraumgemeinschaft*), led by the strongest of them, but which had no right to call the *Großraum* its own *Lebensraum*. For Flohr, 'Eurafrica', as he called it, consisted of the *Lebensraumgemeinschaft* Europe and the *Ergänzungsraum* Africa (1942: 404). In this context it has to be remembered that the rise of geography as an academic discipline and the creation of chairs in geography at the universities between 1860 and 1910 was linked to Germany's national and colonial ambitions (see figure 1).

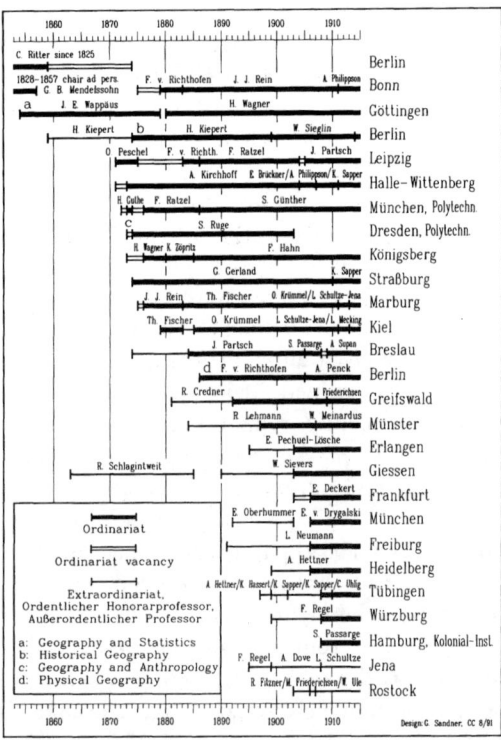

Figure 1   The foundation of geographical chairs at German universities 1860 to 1914 (design: G. Sandner 1991)

A further contribution to the debate was provided by Hans Schrepfer, professor of geography at the University of Würzburg, under the title 'What is meant by *Lebensraum*? A necessary essential definition' (1942). In his article he criticized the usage of the term as highly questionable and wished to limit it to the biological and economic fields. He concluded with the following words:

> German geography today is proud to make its results and its work, which both stand in the service of the entirety of the *Volk* (*Volksganze*), available to the leadership of the state . . . Even more important than pure geography is its application and usefulness . . . But in this way the wish of the geographers to make themselves understood brings out the danger of using scientific language in a popular sense, which cannot attain the level of scientific debate. (Schrepfer 1942: 424)

All these geographers attempted to save the term *Lebensraum* as an academic concept at a time when their colleagues in the field of military and applied geography had a much more practical commitment to fight for German space. This space, in the course of the Second World War, went far beyond the *Lebensraum* concept of the *Volks- und Kulturboden* of earlier years.

Research and Execution of the *Lebensraum* Policy

I would like to draw attention to the way in which the ideological sphere is linked with practical matters, meaning the concrete contributions of geographers to the development of the Nazi state in the Second World War. Although the Nazis did not create a single chair for geopolitics[3] in the occupied eastern territories, and although there was no seminar for geopolitical studies and only a handful of essays on the subject were produced, the state nonetheless had a keen interest in a number of research institutes which included departments of geography (Rössler 1990: 84–111, Burleigh 1988: 253–99).

The main geographical concept used in these institutions was the above-mentioned *Volks- und Kulturbodentheorie*, which had already been developed by prominent geographers during the 1920s. Both the frontiers created by the Treaty of Versailles,[4] and the complex ethnic interrelationships in the lost eastern territories of the German Reich became a central concern within academic research. It persisted after 1933. This research received considerable financial support from the state in both the Weimar Republic and under the Nazi regime. For example, the Nordostdeutsche Forschungsgemeinschaft ('North-East German Research Community', NODFG) was responsible for numerous interdisciplinary studies of ethnic Germandom in both Poland and the Baltic region (Burleigh 1988: 88–9).

In 1943 these 'ethnic German research associations' were incorporated, under the leadership of a geographer from Vienna, Wilfried Krallert, into the Reich Main Security Office (Reichssicherheitshauptamt) of the SS, where they were known collectively as the 'Reich Foundation for Geographical Studies'. These agencies were involved in the complicated evaluation of territories to the east of the German Reich. Concretely, this meant the collection and analysis of population statistics, the calculation of optimal population density, and displays of the assembled material in maps and charts. Studies of this type existed prior to the German invasion of Poland, but with the conquest of ever wider areas in the East this type of research was intensified. The Nazi regime's need for accurate and detailed knowledge of the occupied territories directly benefited this research field: several new research teams and entire institutes came into existence. To mention only one example, the 'East European Research Community' (Osteuropäische Forschungsgemeinschaft) was established in 1942 as an extension to the above-mentioned North-East German Research Community (NODFG). It was particularly concerned with research on ethnic Germandom in Russia and the Ukraine. The different Ethnic German Research Associations (Volksdeutsche Forschungsgemeinschaften; see figure 2, p.67) in turn created so-called Publication Offices, which produced, among other things, detailed maps of ethnic relations in the USSR, or maps showing the proportion of ethnic Germans in the various constituent states of the USSR, including maps of ethnic German villages. These had considerable importance for the planned and partially realized resettlement of ethnic Germans under the slogan 'Home to the Reich' (Rössler/Schleiermacher 1993).

Ethnic geographical research was one of the most important fields of study during both the Weimar Republic and the Third Reich. After the Second World War, however, research in a form so closely compromised by its involvement with the Nazi regime was no longer pursued. In 1947, Carl Troll, professor of geography at the University of Bonn, wrote: 'Today, when the ordering of ethnic relations throughout Central Europe has undergone an ethnic migration of such vast dimensions, a large part of the facts produced by regional and ethnic research has become history with one blow. How will future

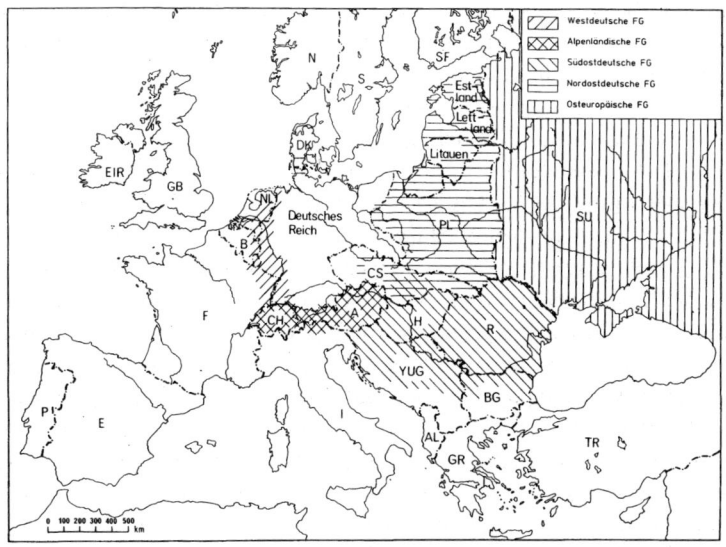

Figure 2   The *"Ethnic German Research Associations"* (Volksdeutsche Forschungsgemeinschaften) *1933–1945 (design: M. Rössler)*

generations regard the published results of such careful research, which was set down on paper just before the great transformation?" (Troll 1947: 17)

Area Research and Regional Planning

The situation in area research and spatial planning was quite different. The Reich Study Group for Area Research (Reichsarbeitsgemeinschaft für Raumforschung) was founded in 1936 under the Four Year Plan, to carry out regional evaluations of the various parts of the *Altreich* (Germany within the borders of 1937). Area research and spatial planning can be considered as a new field and I have described the emergence of a new discipline, which gained ground during the Second World War and after 1945, in great detail (Rössler 1994: 135–7, Rössler 1990: 208–23). In so far as the study of *Raum* (space) was interdisciplinary and concerned with concrete ethnic political goals, it broke new ground and represented a departure from the traditional organization of a discipline like geography in the universities. *Raumforschung* (area research) was soon practised in every German university, and already by 1939 a research project existed, entitled the 'German East'. It had important military ramifications.

As an economist wrote in 1943, *Raumforschung* and *Raumplanung* were confronted with 'mighty tasks' at this time. 'If German *Raum* formation is to measure up to the new economic, logistical, social and demographic, and political objectives, then changes in the composition of towns on a huge scale, the dispersal of major cities and industrial centres, the relocation of factories and the resettlement of people and other transformations will have to be tackled' (Hesse 1943: 89). In view of the far-reaching scope of these plans, it is hardly surprising that the power conflicts, bureaucratic infighting and general inconsistency which was so characteristic of the Nazi regime was in evidence in this area, too. Although officially the Reich Office for Spatial Organization (Reichsstelle für Raumordnung) was responsible for this field of research, its significance was soon attenuated both in regard to its own subordinated research organizations as well as *vis-à-vis* the new research and planning agencies created by Himmler, Rosenberg and the German Labour Front.

One particular theoretical model had already assumed considerable significance in the context of Nazi area planning, and that was the 'central place theory' (Rössler 1989). This theory was based on the dissertation by Walter Christaller, 'Central Places in Southern Germany: An Economic-Geographical Investigation into the Regularities of the Distribution and Development of Settlements with Urban Functions', published in 1933. Essentially, he argued that there was a link between a hierarchically structured network of settlements and a likewise hierarchically structured order of regional economies. His model was dependent upon economic, administrative and political factors.

The application of the central place theory was soon in evidence in the endeavours of the various planning agencies responsible for the occupied East. In 1940, Konrad Meyer summoned Christaller[5] to his own Institute for Agricultural and Political Studies in Berlin, where he was to work with a research team that included geographers, rural sociologists and landscape planners. In this context, Christaller researched the 'cultural and market centres' of the Warthegau, and hence endeavoured to apply his theories to occupied Poland. Significantly, Meyer also employed him as a staff member of the Main Office for Planning and Soil (Hauptamt für Planung und Boden), an agency within Himmler's Reich Commissariat for the Strengthening of

Ethnic Germandom (Reichskommissariat für die Festigung Deutschen Volkstums). Himmler established such a planning office when Hitler entrusted him with the reorganization of ethnographic relations and the creation of settlement areas in the East. This planning agency was also responsible for the overall plan for *Lebensraum* in the occupied East, which later became notorious as the 'General Plan East' (Generalplan Ost). Here, too, one encounters Christaller's theory of a hierarchy of places, referred to in this context as 'settlement pearls', which appear in graduated zones of settlement. Academics on the staff of Meyer's planning office worked on the General Plan East, making calculations concerning cost, available settlers, and the spatial transformation of the area concerned (Rössler/Schleiermacher 1993: 9).

The Institut für Deutsche Ostarbeit (IDO) in Cracow represented a model for German 'cultural' and academic policy in the occupied East (Rössler 1990: 84). The geography section of the IDO was initially concerned with purely descriptive studies of occupied Poland, including a Baedecker tourist guide. Soon, however, it was also involved in planning matters. For example, studies of regional transport and market networks were carried out in the various districts of the *Generalgouvernement*, studies which formed the basis for transport and economic planning with military and logistical effects. Similar institutions were founded at the Reich University of Posen, where there was a Department of Geography, a research programme on 'Ethnic Studies, Frontier and Ethnic Germandom', and a Geographical Research Office of the Warthegau, which was involved with Germanization and settlement planning. A further institution was founded in the occupied USSR: the Institute for Geographical Studies in Kiev, which was concerned with economic and regional geographic studies of the Ukraine (Rössler 1990: 112–33).

As the war went on, numerous geographers were involved in the newly created military geographical agencies: 'MilGeo' for the terrestrial and 'MarGeo' for the sea and coastal zones. From 1943 onwards there was also a specialist for geographical research in the Reich Research Council with various field study groups and research units: the geographer Otto Schulz-Kampfhenkel was nominated by Göring as Sonderbeauftragter für Erdkundliche Forschungen im Reichsforschungsrat (Special Commissioner for Geographical Research in the Reich Research Council) (Rössler 1990: 203).

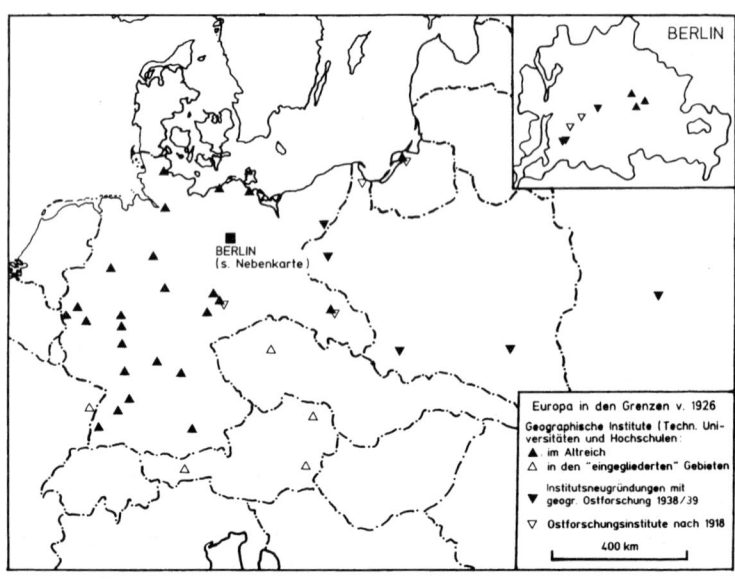

Figure 3   Geographical Research on the East 1918–1945 (design: M. Rössler)

Beyond *Lebensraum*: The Emergence of Colonial Geography and the SS Expeditions

Both living space and colonialism were basic essentials in the imperialist ideology and strategy of the German Reich. However, a number of well-known geographers saw 1933 as a turning point. Hitler gave colonial geography new confidence (Sandner/Rössler 1994: 124) and strong demands were made for new tropical and colonial space. With the Second World War the idea of a new colonial empire gained ground with every victory of the National Socialist army. Colonial geography and the science of colonial planning were seen as sciences of the future. With the support of the Colonial Policy Office (Kolonialpolitisches Amt) of the Nazi Party, geographers embarked on the redistribution of African colonies or on new colonial research projects, including the multi-volume *Afrika-Handbuch* prepared by Erich Obst between 1941 and 1943. Colonial research reached a peak in 1942 when the Nazi Empire became one of the most rapidly expanded land empires in history.

Carl Troll, founder of the concept of landscape ecology, was involved in this research, which was based on new technologies, including aerial photography. His article on 'Aerial plan and ecological soil research. Its potential usefulness for scientific research and the practical exploration of unknown countries' was the starting point for modern aerial photographic analysis, as well as for the notion of 'landscape ecology' (*Landschaftsökologie*), which could be defined as interaction between life and environment (Troll 1939: 297). Troll's research was based on expeditions to East Africa, which were partly financed by the Kaiser-Wilhelm-Institut für Züchtungsforschung (Böhm 1991: 404). He even pointed out that the colonial geography approach would mean no competition with the large eastern and internal settlement planning, as aerial photographic research could be applied to any area or region.[6]

At the same time, Troll was also involved with numerous other geographers and anthropologists in the very popular scientific expeditions which were carried out by the SS Ahnenerbe (Kater 1974; Greve 1995; Rössler/Schleiermacher 1997). The Forschungs- und Lehrgemeinschaft das Ahnenerbe was founded in 1935 to analyse with 'exact scientific methods' the space and heritage of 'Indogermandom'. Himmler became president of the Ahnenerbe in 1939, and a total of thirty-five research institutions and units were created. The expeditions to the Hindu Kush (1935), Nanga Parbat (1934, 1937) and Tibet (1943) were front-page issues in the *Völkischer Beobachter* and raised the popularity of the regime considerably. Carl Troll, who carried out a botanical-geographical profile during the Nanga Parbat expedition, returned to Germany as a national hero. Probably the best-known expedition was the SS Tibet expedition, undertaken in 1943 and led by the geographer Ernst Schaefer. It had clear political (further expansion to the east) and ideological objectives (research on the origins of 'pure wheat', i.e. an original strain of wheat). Schaefer's Film *Secret of Tibet* (*Geheimnis Tibet*) was a cinematic production for the general public in Nazi Germany. Discussions about the SS expeditions began again with the distribution of the film *Seven years in Tibet* (1998, by Jean Jacques Annaud, with Brad Pitt). This film is based on the book by Heinrich Harrer, whose SS involvement was known and well documented (Ziegler 1997). His attitude after 1945, which was to avoid critically reviewing his involvement in Nazi politics and practice, is typical of practically all mountaineers and most geographers in the Third Reich.

What Happened in 1945? Nazi Geographers Employed by American Agencies

A subdivision of the Office of Military Government for Germany (OMGUS) took over the former Abteilung für Landeskunde, which was the division for geographical studies of the Reichsamt für Landesaufnahme ('Reich Office for Land Studies'). The Abteilung für Landeskunde had been founded in April 1941, when topographical material for the occupation of the East in particular was needed (Rössler 1990: 154). It was led by one of the most prominent geographers during the Third Reich, Emil Meynen. He was the managing director of the 'Ethnic Germandom Research Units' in Berlin, and was thus involved in all *Lebensraum* research after 1939. At the end of June 1945, the Abteilung für Landeskunde's division for geographical studies was brought to the little town of Scheinfeld by the American military geographers Lieutenants Thomas R. Smith and Lloyd D. Black (Böhm 1991: 309). Surprisingly, the American administration needed regional and topographical information for the division of Germany. The Americans had, during the Second World War, one of the most important agencies for regional information, the Office for Strategic Services (OSS), where geographers and social scientists cooperated in the so-called Research and Analysis Branch (Rössler 1996). From 1941 to 1945, there were over 3,000 original maps produced, mostly on Europe, as well as hundreds of regional studies. Therefore, in principle, there was no need to employ German geographers for the production of maps, charts, landscape and vegetation maps as well as 'Kreislandeskunden', the geographical description of administrative units. The American officer Black wrote later: 'The Abteilung für Landeskunde . . . represents the greatest actual and potential force in German geography today. It was re-established financially in December 1945, with a mandate to initiate and coordinate geographical research in Germany and has now been organized into eleven subdivisions, each responsible for a certain part of the program' (Black 1947: 148).

For British officials, the whole story was highly dubious. Therefore the EPES (Enemy Personnel Exploitation Section) came to Scheinfeld in August 1946 and imprisoned Meynen and his staff. Together with the staff of Albert Speer, the Stab Osenberg (founder of the Wehrforschungsgemeinschaft) and representatives of I.G. Farben, they were brought to a camp at Kranzberg castle, where they were interrogated. This procedure was called 'Operation Dustbin'. Meynen and his staff were interrogated by the British Major Tilley and forced to write 'homeworks' on their work during National Socialism. One of them is a report on their eastern research, entitled 'Der Drang nach Osten', of over 300 pages.[7] This manuscript, a team work by several geographers, mainly Erich Otremba, Angelika Sievers and Emil Meynen, was a 'whitewash' paper on their research for the *Lebensraum* policy in the Third Reich.

The American agencies helped re-establish the Abteilung für Landeskunde in 1945, a research unit previously involved in the *Lebensraum* policy. The Scheinfeld office was the first step towards establishing one of the main research and planning institutions of the Federal Republic of Germany: the Bundesforschungsanstalt für Landeskunde und Raumordnung led by Emil Meynen for many years afterwards. The differences in the approaches of the British and American agencies have to be noted, and a thorough analysis would be interesting. This example of continuity in personnel can be illustrated by numerous other cases, including that of one of the main actors of the General Plan East, Konrad Meyer. After he left the Nuremberg Trials as a free man, he became professor in landscape planning at Hanover University.[8]

The question of continuity and discontinuity after 1945 is an interesting one, as we can note continuity of personnel and research institutions, as well as discontinuity mainly in the field of military geographical research. A number of research themes were highly compromised: e.g. colonial geography or *Volkstum* research, which stopped completely. Area research and regional planning, as well as landscape ecology, illustrating the 'modernity' of research under the National Socialist system, developed as separate fields in the German universities and in the institutions of both the Federal Republic of Germany and the German Democratic Republic. It was only in the 1980s and 1990s that critical reflection began on the role of geography during the Third Reich and the origins of newly evolving disciplines (area research, spatial and regional planning) so closely related to geography, and only during the last few years has this reflection come from within these disciplines (Schmals 1997; Akademie für Raumforschung und Landesplanung 1996). The former 'collective silence' (*das kollektive Beschweigen*) has finally been broken.

Notes

1. I wish to express my gratitude to Professor Gerhard Sandner for
    providing illustrations to this article and for his support in carrying out
    the research project at the University of Hamburg. Debates in different
    fora and discussion groups were crucial for the findings of my reasearch.
    Among these different discussion groups were 'Geography and Fascism'
    (together with Dominik Siegrist and Michael Fahlbusch), the 'Verein
    zur Erforschung der nationalsozialistischen Gesundheits- und Sozial-
    politik' (Hamburg) and the 'Hamburger Stiftung für Sozialgeschichte'
    (now Bremen). Concerning recent debates on geographers and historians
    during National Socialism see in particular Fahlbusch (1998, 1999),
    Herbert (1998) and Schöttler (1997), as well as for the history of planning
    and area research Reichel (1997), Rössler (1993, 1994) and Münk
    (1993).
2. The translation of the terms is extremely difficult. 'The German
    Volksboden is accompanied by a characteristic German *Kulturboden*, which
    is different from neighboring areas of culture. It is characterized by an
    extremely careful form of cultivation, which does not grind to a halt
    when it encounters difficulties' (Penck 1926, cit. in Burleigh 1988: 26).
3. Geopolitics could be seen as more important during the creation of the
    Nazi imperialist ideology in the Weimar Republic than after 1933.
    Haushofer's influence was largely overestimated, in particular by
    American scholars. Recent studies by Guntram Herb (1997) or David
    Murphy (1997) strengthen this line of argument.
4. The frontiers were created by American and French geographers. The
    Americans hired a ship with thirty geographers under the leadership of
    the very influential Isaiah Bowman and drew the maps for the Paris peace
    conference. On the French side, it was Emanuel de Martonne and his
    team (Mehmel 1995).
5. The biography of Walter Christaller, born 1893, is a very complex one
    (see Rössler 1987). After the completion of his thesis (1933), he was
    involved in the production of the *Atlas des deutschen Lebensraumes*.
    Christaller, who had become a pacifist after the First World War, tried
    to go into exile in France in 1934, but did not succeed to make a living.
    He worked at Freiburg University (1937–40) with Theodor Maunz, the
    founder of the concentration camp law (KZ Recht). In 1940 he joined
    the Reichskommissariat für die Festigung Deutschen Volkstums under
    Himmler. He became a member of the Communist Party after 1945.
6. I would like to refer to the fundamental research carried out by Hans
    Böhm on Carl Troll and his role during National Socialism (Böhm 1991;
    1995). Böhm reviewed the archival material on Troll in the Archive of
    the Geographical Institute in Bonn.
7. Public Record Office Kew/London, FO 1031/140.

8. For a detailed analysis see Mechtild Rössler, 'Konrad Meyer und der Generalplan Ost in der Beurteilung der Nürnberger Prozesse', in Rössler/ Schleiermacher 1993: 356–65.

References

Akademie für Raumforschung und Landesplanung (ed.) (1996), *50 Jahre ARL in Fakten*, Hannover: Akademie für Raumforschung und Landesplanung.

Black, Lloyd D. and Smith, Thomas R. (1946), 'German Geography: War Work and Present Status', *The Geographical Review*, 36: 398–408.

Black Lloyd D. (1947), 'Further Notes on German Geography'. *The Geographical Review*, 37: 147–8.

Böhm, Hans (ed.) (1991), *Beiträge zur Geschichte der Geographie an der Universität Bonn* (Colloquium Geographicum 21), Bonn: F. Dümmlers Verlag.

—— (1995), 'Wissenschaftliche Überwinterung - Angewandte (kriegswichtige) Forschung – Rettung eines Paradigmas', in Ute Wardenga and Ingrid Hönsch (eds), *Kontinuität und Diskontinuität der deutschen Geographie in Umbruchphasen. Studien zur Geschichte der Geographie* (Münstersche geographische Arbeiten 39), Münster: Institut für Geographie der Westfälischen Wilhelms-Universität Münster, 129–40.

Burleigh, Michael (1988), *Germany Turns Eastwards. A Study of Ostforschung in the Third Reich*, Cambridge: Cambridge University Press.

Chickering, Roger (1984), *We Men Who Feel Most German. A Cultural Study of the Pan-German League, 1886–1914*, Boston: Unwin.

Fahlbusch, Michael (1994) *'Wo der deutsche . . . ist, ist Deutschland'. Die Stiftung für Deutsche Volks- und Kulturbodenforschung in Leipzig 1920–1933* (Abhandlungen zur Geschichte der Geowissenschaften und Religion/Umwelt-Forschung, Beiheft 6), Bochum: Brockmeyer.

—— (1998), 'Die "Volksdeutschen Forschungsgemeinschaften": Ein Brain-Trust der NS-Volkstumspolitik?' Paper presented at the 42. Deutscher Historikertag, Frankfurt, 10 September 1998.

—— (1999), *Wissenschaft im Dienst der nationalsozialistischen Politik? Die Volksdeutschen Forschungsgemeinschaften von 1931 bis 1945*, Baden-Baden: Nomos.

——, Rössler, Mechtild and Siegrist, Dominik (1989), *Geographie und Nationalsozialismus. 3 Fallstudien zur Institution Geographie im Deutschen Reich und der Schweiz* (Urbs et regio 51), Kassel.

Flohr, Ernst Friedrich (1942), 'Versuch einer Klärung des Begriffs Lebensraum', *Geographische Zeitschrift*, 48: 393–404.

Greve, Reinhard (1995), 'Tibetforschung im SS Ahnenerbe', in Thomas Hauschild (ed.), *Lebenslust und Fremdenfurcht. Ethnologie im Dritten Reich*, Frankfurt: Suhrkamp, 168–99.

Herb, Guntram H. (1997), *Under the Map of Germany. Nationalism and Propaganda 1918–1945*. London and New York: Routledge.

Herbert, Ulrich (ed.) (1998), *Nationalsozialistische Vernichtungspolitik 1939-1945. Neue Forschungen und Kontroversen*. Frankfurt am Main: Fischer.

Hesse, G. (1943), *Die deutsche Wirtschaft. Volkswirtschaftlehre*, Hannover: Meyer.

Kater, Michael H. (1974), *Das Ahnenerbe der SS 1935–1945. Ein Beitrag zur Kulturpolitik des Dritten Reiches* (Studien zur Zeitgeschichte 6), Stuttgart: dva.

Kuntze, Paul H. (1938), *Das Volksbuch unserer Kolonien*. Leipzig: Dollheimer.

Mehmel, Astrid (1995), 'Deutsche Revisionspolitik in der Geographie nach dem Ersten Weltkrieg', *Geographische Rundschau*, 47: 498–505.

Münk, Dieter (1993), *Die Organisation des Raumes im Nationalsozialismus. Eine soziologische Untersuchung ideologisch fundierter Leitbilder in Architektur, Städtebau und Raumplanung des Dritten Reiches*, Bonn: Pahl-Rugenstein.

Murphy, David M.(1997), *The Heroic Earth: Geopolitical Thought in Weimar Germany, 1919–1933*, Kent, OH: Kent State University Press.

Obst, Erich (1941), 'Ostbewegung und afrikanische Kolonisation als Teilaufgaben abendländischer Großraumpolitik', *Zeitschrift für Erdkunde*, 265–78.

Prigge, Walter (1986), *Zeit, Raum und Architektur. Zur Geschichte der Räume* (Politik und Planung 18), Köln: Deutscher Gemeindeverlag.

Reichel, Peter (1997), *Das Gedächtnis der Stadt. Hamburg im Umgang mit seiner nationalsozialistischen Vergangenheit* (Schriftenreihe der Hamburgischen Kulturstiftung 6), Hamburg: Dölling und Galitz.

Rössler, Mechtild (1983), *Die Geographie an der Universität Freiburg 1933–1945. Ein Beitrag zur Wissenschaftsgeschichte des Faches im Dritten Reich*. Freiburg: Staatsexamensarbeit.

—— (1987), 'Die Institutionalisierung einer neuen "Wissenschaft" im Nationalsozialismus. Raumforschung und Raumordnung 1935–1945', *Geographische Zeitschrift*, 75: 177–94.

—— (1989), 'Applied Geography and Area Research in the Nazi Society: The Central Place Theory and its Implications, 1933 to 1945', *Society and Space*, 7: 419–31.

—— (1990), *Wissenschaft und Lebensraum. Geographische Ostforschung im Nationalsozialismus*, Berlin: Dietrich Reimer.

—— (1994), '"Area Research" and "Spatial Planning" from the Weimar Republic to the German Federal Republic: Creating a Society with a Spatial Order under National Socialism', in Monika Renneberg and Mark Walker (eds), *Science, Technology, and National Socialism*, Cambridge: Cambridge University Press, 126–38.

—— (1996), 'Geographers and Social Scientists in the Office for Strategic Services (OSS) 1941–1945', in Vincent Berdoulay and Hans van Ginkel (eds), *Geography and Professional Practice* (Nederlandse Geografische Studies 206), Utrecht: Koninklijk Nederlands Aardrijkskundig Genootschap, 75–85.

—— and Schleiermacher, Sabine (eds) (1993), *Der 'Generalplan Ost'. Hauptlinien der nationalsozialistischen Planungs- und Vernichtungspolitik*, Berlin: Akademie-Verlag.

—— and Schleiermacher, Sabine (1997), 'Himmlers Imperium auf dem "Dach der Erde": Asien-Expeditionen im Nationalsozialismus', in Michael Hubenstorf et. al. (eds), *Medizingeschichte und Gesellschaftskritik. Festschrift für Gerhard Baader*, Husum: Matthiesen, 436–53.

Sandner, Gerhard and Rössler, Mechtild (1994), 'Geography and Empire in Germany 1871 to 1945', in Anne Godlewska and Neil Smith (eds), *Geography and Empire* (Institute of British Geographers: Special Publications Series 30), Oxford and London: Blackwell, 115–27.

Schmals, Klaus (ed.) (1997), *Vor 50 Jahren . . . auch Raumplanung hat eine Geschichte* (Dortmunder Beiträge zur Raumplanung, Blaue Reihe 80), Dortmund: Informationskreis für Raumplanung.

Schmitthenner, Heinrich (1938), *Lebensräume im Kampf der Kulturen*, Leipzig. Second edition (1951), Heidelberg: Quelle & Meyer.

—— (1942), 'Zum Begriff "Lebensraum"', *Geographische Zeitschrift*, 48: 405–17.

Schöttler, Peter (ed.) (1997), *Geschichtsschreibung als Legitimationswissenschaft 1918–1945*, Frankfurt am Main: Suhrkamp.

Schrepfer, Hans (1942), 'Was heißt Lebensraum? Eine notwendige begriffliche Klärung', *Geographische Zeitschrift*, 48: 417–24.

Smith, Woodruff D. (1986) *The Ideological Origins of Nazi Imperialism*, New York/Oxford: Oxford University Press.

Troll, Carl (1939), 'Luftbildplan und ökologische Bodenforschung. Ihr zweckmässiger Einsatz für die wissenschaftliche Erforschung und praktische Erschliessung wenig bekannter Länder', *Zeitschrift der Gesellschaft für Erdkunde zu Berlin*, 241–98.

—— (1947), 'Die geographische Wissenschaft in Deutschland in den Jahren 1933 bis 1945', *Erdkunde*, 1: 3–48.

Ziegler, Senta (1997), 'Die Nazi Akte Harrer', *NEWS* 23: 220–2.

HELMUTH TRISCHLER

Aeronautical Research under National Socialism: Big Science or Small Science?

Defining concepts is amongst the most difficult and least-loved tasks facing the historian. It leaves him open to the criticism that concepts vary according to time and place. Supporters of 'linguistic turn' even argue that the historiographical formulation of the past is purely linguistic in nature or, even more radically, that in the final analysis reality is only constructed in language (cf. Evans 1997: 191–224; Lorenz 1998). On the other hand, the representatives of history as social science argue forcefully that the historian should make clear his definitions. I want to avoid this dilemma by showing firstly the origins of the term *big science* rather than defining it.

Air transport, like space travel, is one of those scientific and resource-intensive technologies in which large-scale research is evident from a comparatively early stage. I would like to demonstrate this with a German example in the second section of my essay. In the third section I want to pursue the question of how aviation research developed under the conditions of the Second World War. And here I want to ask in particular, with which instruments the Nazi state tried politically to influence this rapidly expanding field of research and how scientists reacted to this attempt at political control. Finally I shall address the question in the title, whether this was 'big science or small science?', and seek a balanced answer from international comparisons.

The 'Big' in Science: Concept and Definition

The search for the roots of big science leads us first of all
to the Manhattan Project as much from a phenomenological as from
a terminological perspective. The research into and the development
and building of the two American atom bombs *Fat Man* and *Little Boy*
swallowed the huge sum of more than $1 billion between 1942 and
1945. In research and testing facilities throughout the United States,
around a quarter of a million workers were employed, requiring a
hitherto unheard-of degree of central planning. Out of the Manhattan
Project a whole range of research establishments emerged after the
end of the Second World War. They are now recognized as national
laboratories: the Military Research Center at Los Alamos, as well as
the nuclear research institutes at Argonne (ANL), Brookhaven (BNL)
and Oak Ridge. A similar development can be seen in European sister
establishments like Harwell, Great Britain, and Saclay, France,
although in Germany Allied dismantling and the ban on research
caused an institutional break. Nuclear research at Karlsruhe and
Jülich was rebuilt from scratch after the end of Allied control.

It was no coincidence that it was scientists at nuclear establishments
who coined the phrase 'big science'. Lew Kowarski, technical director
of the French atomic energy commission (CEA), was analysing the
change in research structure associated with 'large scale physical
research' as early as 1949. Alvin M. Weinberg, director of the national
laboratory in Oak Ridge, made the term 'big science' popular in the
course of the 1960s and with Derek de Solla Price it was adopted for
the social sciences. In Germany the State Secretary at the Federal
Ministry of Research, Wolfgang Cartellieri, introduced 'big science'
as *Großforschung*. This German term triumphed over the alternative
proposal of the nuclear physicist Wolf Häfele, who, after his return
from a position as guest researcher at Oak Ridge in 1959/60, suggested
the term *Projektwissenschaft* for goal-oriented scientific schemes like
the fast-breeder plan he developed in Karlsruhe (Kowarski 1949;
Weinberg 1967; Price 1963; Cartellieri 1963; Häfele 1963; cf. Weiss
1988: 152–3).

At second glance it is clear that the history of big science stretches
back further than the Manhattan Project. It goes back to the secular
process of modernization in science and society. Modern science
evolved not only in the studies of individual scholars, working like
isolated monks on chosen topics, but also more and more in the work
of teams of specialists from different disciplines working on common

resource-intensive projects. The socialization of research manifested itself in the late nineteenth and early twentieth centuries in three institutional innovations which all originated in Germany, whence they were adopted by other industrializing nations. These innovations were: the creation of university laboratories; the development of industrial research; and the evolution of a varied landscape of non-university research institutions.

Germany served as the prime mover in the development of the scientifically based national systems of innovation which appeared at the turn of the century (Nelson 1993). The dynamic nature of the change was clear to contemporaries. The historian Theodor Mommsen felt in 1890 that science also had a social dimension: 'As with the large state and with large industry, so large science, which is led by individuals rather than performed by them, is a critical element of our cultural development, as are the academics who are involved in it, or at least they should be.' The theologian and later President of the Kaiser-Wilhelm-Gesellschaft, Adolf von Harnack, echoed Mommsen in 1905 when he spoke of the team-work of teachers at local, national and international levels, whether in the production of encyclopaedias, or in natural sciences like meteorology or astronomy (Ritter 1992: 13).

The First World War can be viewed as a phase of accelerated historical change. In the course of the four years of the Great War, science was effectively transformed in regard to both its content and its socio-political significance. In the midst of a fundamental crisis of national security, those nations ready for war learned to perceive science and technology as an essential resource and to concentrate scientific resources on specific projects. In order to use these resources more efficiently, new systems of control were created, linking science, technology, the economy and government. These systems proved to be as stable as the institutions in which large scientific projects were conducted. They outlived the war as well as the subsequent demobilization and served as spearheads for a new expansion of science for wars to come. Without resorting to the stereotype of 'war as the father of all things', it is clear that the First World War, and even more so the Second World War, can be seen as important transition stages in the development of big science (cf. Trischler 1996). This case study on aeronautical research, a science of great military importance, will provide evidence for this development, which also expresses the newly forged connection between science and society. And it will show us its dialectical development, which is of long-term historical importance and can be described as a part of the overall modernization process in the twentieth century.

Big science has a long history. It has roots in developments which go right back in early modern history. These include secular trends like the increasing impact of science on society; the growing inter-dependence of science and technology; the increasing expenditure on personnel and apparatus; as well as the importance of research results and their relevance for the science-based regulatory state, with its growing expenditure on administration and standardization. In the incubation of big science, which I date from 1914 to 1945, several new initiatives gave this type of research an institutional form. It had the following characteristics:

-   the bringing together of different scientific disciplines in projects involving large apparatus;
-   the combination of various resources and personnel;
-   overwhelming state finance;
-   the setting of concrete middle- and long-term goals;
-   the setting of goals that are claimed to be of special political and social relevance;
-   a conflict between political goal-setting and the ongoing autonomy of scientists in the setting of concrete objectives.

Big science is characterized by size, by which one means expansion in several directions: geographic in its extension over wide areas and in its scientific and economical impact on whole regions; economic in its execution of projects that consume billions; technological in its grouping around big, sophisticated instruments; organizational in the size and multidisciplinary nature of the groups involved; functional in the concentration on concrete, large projects; and frequently also international in the collaboration of scientists who differ in language, training, research style and cultural background.

The criterion of sheer size is, however, not sufficient to distinguish big science from small science. Big science is not simply big in the quantitative sense. What is specific to big science lies far more in the degree of interplay between three aspects of society: the state, the economy and science. Big science is directed at goals which are held to be political or social priorities. Big science is aimed at mobilizing scientific knowledge for a technological purpose which will benefit the economy, often in connection with a nation's welfare and security interests. It is the state, to an overwhelming degree, and industry to a lesser degree, which finances big science. As a result of this location in the middle of the triangle of state, economic and scientific interests there arises an element of uncertainty. Big science is at the centre of conflicts between competing interests within society (cf. Galison/ Hevly 1992; Szöllösi-Janze/Trischler 1990: 13–14; Trischler 1995: 112–13).

Big Science in the Making: Aeronautical Research in Germany 1900–1940

Aeronautical research was established as an institution-alized science in the years between the turn of the century and the First World War. In 1904, the experimental physicist Ludwig Prandtl - later called the 'father of aerodynamics' – presented his discovery of the boundary layer theory to a scientific audience. More and more anomalies had become apparent in fluid dynamic theory in the second half of the nineteenth century. Traditional theory could offer little more to the technicians in fluid dynamics. Prandtl opened the discipline up to new research in aerodynamics and with his explan-ation of turbulence he created an entirely new research field.

His work was as applicable to fluid dynamics as it was to aero-dynamics. There was no scientific reason for Prandtl and his colleagues to concentrate on aerodynamics; it was rather political reasons that made aerodynamics the central focus of aerospace research. It was the demand from the infant civil and military aviation branch that forced the pace of science. On the eve of the First World War, there were four indicators of the future (Mehrtens 1996: 92–5; Rotta 1990; Trischler 1992: 34–88):

- the foundation of the Motorluftschiff-Studiengesellschaft (Society for the Study of Motor Airships) in 1907, which one year later established an Experimental Institute. The institute served as the birthplace of the world-famous Aerodynamische Versuchs-anstalt (Aerodynamical Research Establishment, henceforth AVA) in Göttingen during the First World War;
- the foundation of the Deutsche Versuchsanstalt für Luftfahrt (German Research Centre for Aviation, henceforth DVL) in 1912;
- the formation of a personal network of aeronautical experts from state, science and industry in the Wissenschaftliche Gesellschaft für Luftfahrt (Scientific Society for Aviation, henceforth WGL);

- the institutionalization of aeronautics within the curricula of major German universities.

England and France followed the same pattern. The sharpening tensions within Europe forced the Entente to follow suit and fund the rapidly expanding aviation sector (Walker 1971; Chadeau 1985). The national crisis of the war accelerated the establishment of research institutions and structures. Everywhere aviation research was expanding. Politicians and soldiers realized more clearly than before the importance of science for the development of weapons of war. Metal aircraft construction, exhaust-driven superchargers and balloon bombers showed that the way from basic research to technical application could be drastically shortened by the concentration of resources (cf. Ziegler 1994).

Other developments in aeronautical research in Germany were deeply influenced by the political situation in Europe. The expansionist policy of the war had been seriously curtailed by the Treaty of Versailles. Military aviation was completely forbidden, and civil aviation was severely limited in its possibilities until the mid-1920s. This resulted in industry's refusal to support research programmes and left the sciences to fight for survival during the inflationary period between 1918 and 1923. However, the inflation proved to be of great importance for science. The devaluation of the capital of private research organizations forced the state to come to their aid. From the beginning of the 1920s, the Reich placed annual subsidies at the disposal of the Kaiser-Wilhelm-Gesellschaft as well as the DVL. Aeronautical sciences, supported financially by the Reich, were revived during the inflation crisis and were thus able to make plans for the future (Witt 1990; Trischler 1992: 120-34).

Prandtl had for a long time planned to found an institute for aerodynamics attached to the Kaiser-Wilhelm-Gesellschaft which would work alongside the AVA. After long negotiations in the spring of 1924, immediately after the currency reform, building began on the Kaiser-Wilhelm-Institut für Strömungsforschung (Kaiser-Wilhelm-Institute for Hydrodynamics). The institute began work in July 1925. Göttingen developed in the 1920s into a much-visited international Mecca for aerodynamic research. The institute which was directed by Ludwig Prandtl and Albert Betz became the nursery for similar aerodynamic research institutes all over the world. Göttingen was also at the head of conceptual research work. The Göttingen wind tunnel with its closed wind stream proved itself to be a superior instrument for testing scientific theories through experimental methods and led to many technical innovations. Prandtl's airfoil theory, a general theory of lift and drag published in 1918, and the pathbreaking investigations into axial-flow compressors by Albert Betz and Walter Encke during the 1920s and 1930s, show the fertility of Göttingen's system (Hanle 1982: 23–52; Constant 1980: 99–116; Rotta 1990).

Peter Fritzsche (1992) has rightly characterized the German public of the 1920s as strongly air-minded. Precisely because of the Allied restrictions on the building of aircraft, German aviation technology became a symbol of national greatness and a compensation for the loss of national sovereignty resulting from the Treaty of Versailles. It was for this reason that Charles Lindbergh's single-handed crossing of the Atlantic in 1927 was a sobering experience for the Germans, for it seemed to demonstrate the technological superiority of the Americans. German scientists believed that an important reason for the American strength in aviation lay in the fact that the National Advisory Committee for Aeronautics (NACA) was directly under the responsibility of the President. They wanted a similarly direct control system for German research. Even those scientists who felt that these high-flying plans were unrealistic, conceded that in Germany there was no effective coordination of aviation research. In Stuttgart, Darmstadt, Munich and elsewhere new research institutions had been set up which were working autonomously, and there was no sign of coordination in research and industry. In order to use financial resources more effectively, the Reichsverkehrsministerium (Reich Transport Ministry) proposed a cooperative research programme under the leadership of the quasi-state-run DVL. The scientists, however, were worried about their independence and proposed an alternative. They offered to set up a commission which would work in close cooperation with the ministry to set up projects across individual institutes, determine the work programmes of the institutes and coordinate the research with the development work in industry (Trischler 1992c: 145–50).

Leaving this political experiment to the self-regulation of the scientists paid off for the government. The Deutscher Forschungsrat für Luftfahrt (German Research Council for Aviation), which had been set up in 1928, succeeded in improving the academic agreement among scientists, as well as coupling the scientists' goals to those of the state and industry. It was important that the Notgemeinschaft der Deutschen Wissenschaft (Emergency Association in Aid of

German Science), a state-financed organization founded in 1920 to support academic research, was involved in this new definition of the relationship between state, science and industry. Research on high-altitude flying was one of the priorities of the Notgemeinschaft (cf. Marsch 1994).

Just before the Nazi seizure of power a corporatist research policy was evident in aviation. All three parts of the research system pursued their own interests. This corporatist research policy is all the more remarkable as at the same time the presidential regime and the rise of National Socialism were wearing away the basis of political corporatism in its liberal variant (cf. Abelshauser 1984). The world economic depression hit the aviation industry particularly hard. Those firms with a weak capital base, which were dependent on demand from the state, had only survived the 1920s thanks to generous public subsidies. When these were cut back as part of the deflation policy of governments after 1929, many companies faced bankruptcy (cf. Budraß 1998: 273–91). The cuts in state development programmes also affected aviation research. The ambitious expansion programme of the DVL was a victim of the Reichssparkommissar (Reich Savings Minister) cuts. The AVA also suffered under the financial crisis. It had to do without the large 6-metre-diameter wind tunnel which had been planned since the middle of the 1920s. However, the idea that aeronautical research was robbed of its chance to develop is exaggerated. In comparison to the existential crisis that was facing industry, research survived the crisis relatively unscathed. Reich subsidies only sank in 1932/3 to 75 per cent of their 1928/9 level. Up to 1931 the level of personnel was actually increased and thereafter redundancies were kept below average (Trischler 1992c: 161–9, and for the following: ibid. 174–206).

However, the scientists perceived the policies of the Reich government in a completely different way. They got the impression that the parliamentary democratic state was generally not in a position to meet their demands. Parliament and state bureaucracy seemed to be unable to see the necessity of supporting an improvement of research installations and facilities. Scientists generally tend to judge themselves against their colleagues at home and abroad. The German aeronautical science community looked to America, which since the 1920s had been a shining example of well-equipped and organized research. The German scientists were forced to sit in silence while in Great Britain, France, and particularly in the United States the foundations for excellent research opportunities were laid, while in

Germany, working with obsolete equipment, it was hardly possible to conduct model tests on new aircraft types. In consequence, and as a reflection of what was happening in German society as a whole, scientists ceased to accept the Weimar Republic as a valid form of government and began looking for alternatives.

Thus, the destruction of parliamentary democracy by the National Socialists was largely well received in the aeronautical scientific community. With undisguised satisfaction the scientists noted that aeronautics was granted autonomy in the new Third Reich. With the appointments of the former director of Lufthansa, Erhard Milch, to undersecretary and Adolf Baeumker as the head of the research department of the newly formed Reichsluftfahrtministerium (Reich Aviation Ministry, henceforth RLM), hopes increased that the importance of research would finally be recognized by the state. Baeumker had gained the trust of the scientists in the 1920s as the official responsible for aviation research in the Reichsverkehrs-ministerium (Reich Transport Ministry). The son of a well-known Munich philosophy professor, he seemed to guarantee the autonomy of science and an unbureaucratic approach to new ways of organizing research.

Prandtl and his colleagues were not to be disappointed. Only weeks after the Nazi seizure of power, the AVA got permission to construct the large wind tunnel, a request the centre had petitioned for in vain for almost a decade. The increase of the budget for aviation by more than 40 million Reichsmark from the job-creation programme made it possible for the DVL to carry out plans for expansion which had been gathering dust since the late 1920s. Aeronautical scientists soon discovered that even their most outrageous demands were met. Research installations which had previously been unthinkable were suddenly approved without question. The financing problem, which had always been the limiting factor of research, no longer seemed relevant.

The expansion of the DVL alone consumed over 28 million Reichs-mark by the beginning of the Second World War, a huge amount, inconceivable by the standards of the Weimar era. At the outbreak of the war the institute had highly modern research facilities, of which only two of the most spectacular need to be mentioned here. The big wind tunnel opened in 1934, had eliptical dimensions of 5 × 7 or 6 × 8 metres and enabled coolers, transmission, propellors and engine casings of large dimension to be tested. Another technological innovation was the *Trudelwindkanal* of 1934/5 which was shaped like

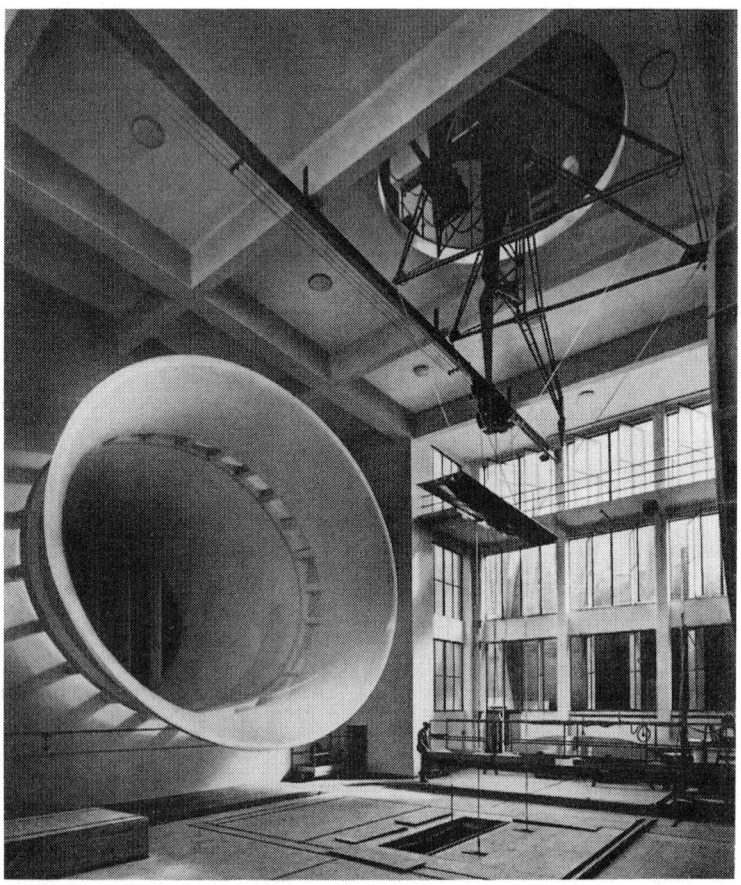

Figure 4   Jet blower of DVL's big wind tunnel.

an enormous egg. In a vertically rising air stream of 4 metres in diameter and 40 m/s hung a model in free movement in front of a camera. The huge dimensions of this wind tunnel were trumpeted by Nazi propaganda. This was the expression of a sort of technological romanticism and the production ethic of National Socialism (cf. Rabinbach 1976; Friemert 1980). The staff of the centre increased threefold within a two-year period. On the eve of the Second World War, the centre had almost 2,000 employees, an expansion in personnel which strained its internal structure. In 1936, the facility was expanded both horizontally and vertically. Between the management and the

Figure 5   DVL's big wind tunnel from outside.

departments new intermediate hierarchical levels were installed. The
autonomy of the departments was cut back, thus enabling the DVL
to take on larger projects, so that there was an improvement in the
quality as well as the quantity of research.

The AVA in Göttingen expanded just as rapidly. Before the Second
World War it looked like a building site. Its new wind tunnel was so
enormous that Lufthansa and Luftwaffe pilots used it as an aid to
navigation. Hardly was the first cold tunnel for testing icing on high-
flying aircraft ready than work began on an even bigger icing tunnel.
In this tunnel an altitude temperature of minus 60 degrees Celsius
und 0.1 bar pressure could be simulated. Its insulation required the
entire annual Portuguese cork harvest (Wüst 1982: 33–4). With the
purchase of the nearby disused limestone quarry and aircraft hangars
including testing equipment, the centre spread right across the middle
of Göttingen.

In 1937, after a long, acrimonious debate, the RLM and the Kaiser-
Wilhelm-Gesellschaft agreed to make the AVA independent and
separate from the Kaiser-Wilhelm-Institut für Strömungsforschung.
As a terminological compromise the name Aerodynamische Versuchsanstalt in der Kaiser-Wilhelm-Gesellschaft was adopted. In return
for generous financial support from the RLM, the centre now had to
work exclusively on aeronautics. The staff grew from 80 employees
in 1933 to over 450 by 1936, and to approximately 700 in the last
year of peace. Albert Betz had to admit that he could no longer run
such a rapidly expanding research concern and in 1939 a separate
administrator began to work at his side. Within half a year the AVA

Figure 6   DVL, Trudelwindkanal: *technological innovation and national socialist aesthetic romanticism.*

had changed fundamentally. Out of a straightforward institute of the Kaiser-Wilhelm-Gesellschaft there had grown a varied and complex research undertaking. Highly modern research facilities were being used or built. In order to be able to handle the rush of orders from the aircraft industry, the wind tunnels were being used in shifts around the clock. Like the DVL, the AVA corresponded to a large degree in size, structure and working methods to the criteria by which we judge big science. The ministry withdrew the scientific head, Albert Betz, from the administration and replaced him with someone they trusted.

The 'great scientific expansion' of the pre-war period (Simon 1947: 24) remained decentralized. Besides the expansion of existing centres, new centres were planned in the mid-1930s. In March 1935, with the proclamation of German air sovereignty, the Nazi government stopped pretending it had no air force (Luftwaffe) and thereby broke the bonds of the Versailles Treaty. The Air Ministry dictated the goal to be attained: Göring's insistence that 'German aeronautical research will have to reach the production levels of the leading foreign nations at the latest by 1938 and then take the lead in several important areas' gave the research department of the RLM new room for manoeuvre. In the internal struggles for power and influence as well as in the negotiations with experts from the military, industry and science, and with competing departments of the polycratic regime, Göring's stated goal was used as a trump card. The personal support of the second most powerful man in the Nazi regime overcame all those obstacles which faced the research department (Baeumker 1944: 31).

Decentralization remained the characteristic of German aviation research. The effort of the DVL to concentrate everything except the

Figure 7　View of Deutsche Versuchsanstalt für Luftfahrt *(DVL), ca. 1935; at the left: the big wind tunnel and the* Trudelwindkanal.

AVA in Berlin-Adlershof would have had many advantages. The building of completely new centres absorbed resources and energies which might have been more effectively used by concentration. The Air Ministry had other concerns, however. The AVA and DVL were reaching their physical limits and could not be protected from enemy air attacks within cities like Berlin or Göttingen. The DVL was anyway considered to be too big to guarantee effective research. A 'healthy decentralization of research across the whole Reich territory' would allow cooperation with regional industries and the full exploitation of personnel resources (Baeumker 1944: 43–4). Hence new establishments were set up, among them the Deutsche Forschungsanstalt für Luftfahrt (German Research Establishment for Aviation, henceforth DFL).

Apart from the building of new centres there was a second model for institutional growth. An existing group of researchers could be taken as the core around which a diversified institute was built. A third variant appeared after 1939. Thanks to the *Blitzkriege*, Germany gained control of important foreign research centres. The potential of these establishments, among them the Etablissement d'Expériences Techniques des Chalais Meudon near Paris, which housed Europe's largest wind tunnel, was channelled into the research landscape of Nazi Germany.

The biggest and most important of these new or extended centres was the DFL. Whereas the DVL was devoted to applied research, the DFL was planned as a centre for basic research. A huge research centre shot up on the green fields outside Brunswick. Wind tunnels of various sizes in the classical Göttingen design were added to research instruments that had hitherto not been able to measure the parameters and phenomena in ballistics and aerodynamics. A cross-section wind tunnel, for example, allowed the study of the influence of side winds up to a speed of 200 m/s on flight to be tested (Blenk 1941: 465). Within just a few short years, the DFL grew into an array of highly advanced laboratories and facilities. Adolf Baeumker, head of the Air Ministry's research department, hit the nail on the head, when in 1942 he stated that the DFL was the largest research project so far realized in Germany (cit. after Trischler 1992a: 174).

The foundation, building and running of the centre in Brunswick were the basic model of research policy in Nazi Germany. The Air Ministry set the long-term goals: accelerated basic research in those areas of use to military aviation like high-speed engine research and weapons. The building of the research centres showed obsessive concern for secrecy and protection against air raids. The setting of the scientific goals, however, was the responsibility of the scientists. The fact that the state controlled the organization of science but not its actual processes meant that scientists enjoyed a high degree of autonomy. The precondition for this cooperation between state and research was the readiness of the scientists to go along with the general political line of the national regime. In fact an analysis of the research work at the DFL shows the high degree to which this new centre fitted in with the rearmament aims of the regime (Trischler 1992c: 213–22).

Even more impressive than the DFL were the regime's plans for the Luftfahrtforschungsanstalt München (Munich Aeronautical Research Establishment, henceforth LFM). As the Stuttgart aircraft firm of Ernst Heinkel began work on jet engines in 1936, research into engines looked like being taken over by industry. The research centre in Brunswick was still under construction and thus was not able to realize its function of producing new ideas and technical innovations in this revolutionary area of aircraft construction. In the Air Ministry, plans were drawn up for a new research establishment in the south of Germany which was to be dedicated to basic research into jet engines. After the *Anschluß* with Austria, Munich was chosen as the site. The nearby Ötztal with its natural resource of water power offered favourable conditions for the planned high-speed wind tunnel with a power of 75,000 kW, in which tests on high-performance engines up to flight speeds could be carried out. With tunnels of this size the Nazi regime hoped to compensate for the apparent superiority of the United States in this strategically important technology. But with the outbreak of the war the building of the Munich centre was postponed. In mid-1940, however, the American Congress passed legislation to encourage engine research. The Nazi regime, despite a shortage of capacity, was determined to catch up. New high-speed tunnels with 8 m diameter as well as test beds for rocket and jet engines were conceived and the building of the facility began. Although the construction of the LFM at Ottobrunn near Munich and Ötztal took the lion's share of the available funds after 1941, the most important projects did not get beyond the basic construction stage before the end of the war. Like a giant shadow – a relic of Nazi giganticism – pieces of the large apparatus stuck out of the idyllic world of the Ötztal (Trischler 1992: 262–9; Hansen 1987: 187–217).

What a difference it made, after the apparent lack of interest by the parliamentary democracy in the necessity of research, to be confronted with the generosity of the new government! It is thus not surprising that the scientists did not regret the passing of democracy and that they quickly aligned themselves with the new dictatorship, particularly when, with regard to their actual work, almost no limits were set to their traditional autonomy. Aeronautics, representing part of the psychological preparation for war, was also allotted a prominent place in school curricula (Fritzsche 1992: 200–3). But nothing can better exemplify the spectacular increase in the power and importance of research in aviation than the foundation of the Deutsche Akademie der Luftfahrtforschung (German Academy of Aeronautical Research) in 1936. Playing up to the well-known vanity of the Reich Air Minister, the head of the Research Department, Adolf Baeumker, succeeded in winning Göring over to the idea of founding an academy for engineering sciences. Göring was very excited and could not wait to follow in the footsteps of the Great Elector, the founder of the Berlin Academy of Science in 1700. Despite violent opposition from the established academies, an academy of technical sciences was created, a novelty in the German research system.

At this stage we can draw together some threads from our discussion. Within just a few years, the policies surrounding research and development had undergone a revolutionary change. In the course of rearmament and preparation for war, the Nazi regime made generous funds available which gave the whole landscape of aviation a new momentum. For aeronautical research the 1930s were without any doubt a period of accelerated change. With reference to our earlier definitions for big science, the following observations can be made:

1. The degree of multidisciplinary work increased. Research areas like gas dynamics or aeroelastics enlarged the disciplines around the central area of aerodynamics. The cooperation of teams of scientists, engineers and technicians became regular practice. Specialization and functional differentiation increased.
2. Increasingly large apparatuses determined the research process. Derek de Solla Price has stated that the law of exponential growth determines the parameters of big science (Price 1963). In aviation research the trend to large-scale growth is clearly visible in the case of the wind tunnel as the most important research tool. Up to the 1950s the power of these tunnels was increased by a factor of 10 over a period of twelve years before stagnation set in for technical reasons (Heinzerling 1990: 320). Wind tunnels, with their visible size and intensity of noise, determined the everyday

work of the research establishment. As the example of the research establishment in Munich shows, they also determined its location. In order to obtain the necessary energy, the Munich centre had to place its high-performance tunnels in the scientific diaspora of the Ötztal. The growing dependence of the research process on big apparatus not surprisingly found little favour amongst traditionally trained scientists. Albert Betz, the head of the AVA, warned in the 1950s of the danger of a 'mechanization of research' through 'intolerably' large apparatus and of the risk that 'real intellectual ability' would be sidelined, referring to his experience in the Third Reich (Betz 1949: 253; Betz 1957: 387).

3. Expanding aviation research consumed a high quantity of resources. The factor which limited growth was, however, not money, as in the Weimar Republic, but a shortage of qualified personnel. Certainly academic aeronautics grew in 1935 by the expansion of so-called *Luftfahrtlehrzentren* (Aviation Academic Education Centres) at the technical universities in Berlin, Brunswick and Stuttgart, as well as partial expansion at Aachen, Darmstadt, Dresden and Munich. However, the universities could not come close to meeting the personnel needs of research. In 1937 a mere sixteen aeronautical engineers received their degrees, and in the following two years only fifty-six and fifty-seven. When one thinks that the RLM needed an annual increase of 3,600 engineers, this was little more than a drop in the ocean (Gundler 1995; Ludwig 1974: 271–83; Homze 1976: 214). During the war the personnel shortage got worse. A further limiting factor was the shortage of construction capacity for building projects. In this regard, the Third Reich showed its inflexibility. Totalitarian regimes are normally seen as infrastructure-friendly. But the sheer range of projects that were undertaken simultaneously in the polycratic chaos of the Nazi regime produced a struggle for resources.

4. Aeronautical research was strongly focused in the 1930s on goals that were considered politically important: the strategic areas of high-altitude flight, jet engines, high-speed aerodynamics, high-frequency research, ballistics and rocket research were systematically improved. Likewise gas turbines, swept wings, tailless aircraft, radar, radio navigation and rocket technology all became development projects with high priority.

5. Aviation research during the Third Reich was to a high degree dependent on the state. The separation of the AVA from the

Kaiser-Wilhelm-Gesellschaft in 1937 shows how the Nazis used both the traditional channels of financial subsidy and other more direct methods to support and encourage their objectives (cf. Macrakis 1993). During the war the regime tried to set the agenda not only for the long-term objectives but also for the details of research.

During the Second World War scientific research was caught between the need to serve politically identified goals and the need to retain scientific autonomy. An analysis of the effect of this process will allow us to give an answer to the opening question: was this big science or small science?

Between State Control and Self-Administration: Aeronautical Research in the Second World War

In contrast to Great Britain, where scientists and technicians were already working on war-related research before the war, the Nazi regime took no concrete steps to prepare research for the forthcoming war. In 1936 the RLM had considered gearing research to the needs of the coming war, but the recommendations of the leading scientists concerning the military relevance of research work were not taken up. The AVA head, Albert Betz, reckoned that in the case of a war two to three times the number of personnel would be required. Instead, aviation research had to surrender half its manpower by January 1940. Those scientists 'without recognizable objectives' were sent to the front or transferred to industry (Ludwig 1974: 289; cf. Boog 1982: 71).

After the outbreak of the war the regime abandoned its long-term research projects in favour of a few pressing war objectives. Research was to concentrate on projects that could be of use to the *Blitzkrieg*. The government determined the budgets of the institutes and insisted that the scientists keep to their written research plans. The RLM had the right to terminate research projects and used this right frequently. Often the work of the centres was interrupted and projects that were considered to be most interesting by the scientists were stopped. The AVA for example had to develop anti-barrage-balloon devices in preparation for the air raids on England and produce the equipment itself. In addition, the regime of secrecy was drastically intensified. The scientists complained that their autonomy was reduced to an extent they had never experienced before.

The longer the war lasted the more the scientists distrusted and disliked the intervention of the military and state bureaucracy in their work. Under the chaotic leadership of Ernst Udet, like Göring a highly decorated and most prominent fighter pilot of the First World War, research lost more and more of its influence within the huge *Technisches Amt* (Technical Office) of the RLM. Ludwig Prandtl's complaint about the 'pure stupidity' of the bureaucrats expresses the desperation of the scientists at being decoupled both from the front and from industry (Trischler 1992c: 246).

After Udet committed suicide in November 1941, the inherent weaknesses of the German war effort in aviation could no longer be denied. Aircraft production, at fewer than 1,000 planes per month, had fallen well below the planned output target. The father of reform for aviation armament, Erhard Milch, now held in his hand the trump card which could break the technological wilfulness of the engineers and entrepreneurs and force them to concentrate armament production on a few standard types (cf. Budraß 1998: 654–703).

At the same time, it became clear that the political attempts to channel aeronautical research had failed. In place of the project-oriented 'big science' the historian is faced with a plethora of isolated 'small sciences'. The diversity of these thousands of scientists' work makes generalization impossible. There was individual resistance to the channelling of working goals according to the needs of the war. In particular those involved in basic research, such as the expert for the boundary layer theory, Hermann Schlichting, were incensed by the limits placed on their autonomy. The need for secrecy undermined communication between colleagues and drove many scientists into a sort of self-imposed isolation. With the growing disillusionment regarding the chances of ending the war victoriously, an increasing sense of vexation prevailed. This vexation often led scientists to a renewed absorption in their work, as they literally buried themselves in their individual projects. But more often than not, scientists were frustrated by the lack of recognition for their willingness to mobilize all their talents for the nation at war. People like Georg Madelung, who were particularly active in the area of the practical applications of research, complained about the lack of contact with the front. He used the example of the First World War to prove his point. Madelung surmised that the reason aeronautical research between 1914 and 1918 had been so successful was that scientists had been able to work in a realistic and practice-oriented environment and had an intensive communication with the front. In contrast the Nazi regime had ignored their requests for contact with industry and with the military (Trischler 1994: 84).

The duplication and overlapping of tasks and a lack of coordination between research and industry was widespread. With the exception of aerodynamics, there was a general lack of cooperation between different research centres. In the critical sectors of arms manufacture such as engines and weapons research there were serious problems. The chaos caused by overlapping projects is illustrated by the fact that by the middle of the war there were around 2,200 ongoing individual projects, not including direct orders from industry. The regime was forced to admit that its attempt to control the course of scientific research had produced the opposite result to that which was intended, namely the concentration of resources on a few short-term projects. Under the pressure to work for politically defined projects, many scientists claimed their research was relevant to the war effort without actually altering the research itself. The increase in resources had led only partly to the concentration of research on big science projects. Often the resources were spread over many small projects carried out by individual scientists without coordination from above.

Again it is clear that sheer growth in size of research undertaken is not in itself a sufficient criterion with which to define 'big science'. The organizational, financial and scientific processes had developed in such a way in the 1930s that aviation research could become big science. However, *de facto* at the beginning of the 1940s 'small science' was still widespread. The main reason for the difference between this planned structure and the actual historical development of research can be found in the changing relationship between the state, industry and science. The traditionally conservative bureaucracy of the RLM was fully aware of the importance of scientific knowledge for the setting of research goals. But its influence was reduced under the pressure of war when research was increasingly forced to meet the military priorities set by the technocrats of the RLM under the influence of Ernst Udet.

The concept of state-funded research was based on the idea of a 'linear model', by which the results of scientific work were directly channelled into industrial development and finally into mass production. Instead of a linear transfer of information between science, technology and industry, in reality the innovation process takes place through heterogeneous channels in which scientific and technical knowledge is mixed up. Complex innovations like the jet engine require a complex relationship between basic and applied research, experimental development, production and use. Derek de Solla Price has aptly characterized this relationship between science and technology as one of 'dancing partners', in which each adjusts his movement to that of the other, sometimes fast, sometimes slow, in such as way that it is impossible to see who is leading whom (Price 1965: 553; cf. Rip 1992).

The concept of the 'linear model' remains in wide use today and it is accepted that there must be a cooperative relationship between research, development and production in order for innovations to be successful. It would be ahistorical to accuse those responsible in the Third Reich for failing to see the complexity of the innovation process. Furthermore, the obsessive secrecy of the regime robbed the scientists of the chance to come into contact with industrial and military partners in order to coordinate the innovation process. To stay with Price's characterization, the researchers imagined themselves to be shabbily treated. They saw themselves condemned to watch helplessly as industry and the military danced in circles around them.

Ultimately however, there was little about the relationship between industry and the military which suggested harmony and synchronization. The aircraft manufacturers had in the course of the war become increasingly distant from the military, whose rapid changes of plan as regards the aims of armaments production made systematic development and production difficult. The polycratic nature of the power relationships in the Nazi regime condemned to failure every attempt to build bridges over the bureaucratic structures which divided science, technology and the military. Polycracy meant isolation as the war progressed and a permanent engagement in a Darwinian struggle for power and resources.

After the entry of the United States into the war and the failure of the offensive on the Eastern Front, the regime changed direction. The Nazi leadership admitted their failure and abandoned the idea that research and technology could work to their command. Returning to the authoritarian traditions of political corporatism, Milch used the corporations – especially the Reichsverband für die deutsche Luftfahrt – to reorganize the war economy, thus circumventing powerful industrialists like Willy Messerschmitt and Ernst Heinkel. On the one hand, industrial self-administration was strengthened and a large part of the responsibility for planning returned to industry.

On the other hand, the regime still controlled the fundamental decision-making process, as can be seen with the massive deflexibilization of aircraft production carried out by Milch and Reich Armament Minister Albert Speer in 1944 (Budraß 1998: 691, 747).

In science, too, the regime returned responsibility to the scientists. The head of the research department at the RLM, Adolf Baeumker, called for a return to the self-governance which had been introduced in the late 1920s. It was fortunate for Baeumker that the issue of scientific policy came onto the agenda in the RLM in the middle of the war. In January 1942 Hitler gave Göring the assignment to reform thoroughly the inefficient Reichsforschungsrat (Reich Research Council), which had been set up in 1936, and to build up a dynamic institution for directing research. According to Göring's bombastic words the new organization was to be as revolutionary in the field of scientific research as the Four Year Plan of 1936 had been for the economy. Baeumker hoped that Göring would not leave 'his' aviation research in the lurch. In fact, aviation research received special treatment, although it was different from that envisaged by Baeumker. The RLM was not willing to give up its research to an external committee, even if this met under the chairmanship of Göring himself. Aeronautical research continued to follow its own course under the reformed system. What did this reform look like and how did it work?

At the centre stood the direction of resources and projects by a powerful leadership figure from the ranks of the scientific community. In Baeumker's words, it was important 'that a single figure stood at the head of aeronautical research who was able to give advice to all the branches of this extensive area and who commanded undisputed respect from the highest Reich offices as well as the research and other institutions' (Trischler 1992c: 248, for what follows: 246–60). The organizational radicalism of his suggestion is clear from the fact that Baeumker was ready to dissolve his ministerial department and place himself under the new leadership.

The leading scientists unreservedly supported this concept and accepted Baeumker's nominee as leader. He had selected the head of the Deutsche Forschungsanstalt für Segelflug (German Research Establishment for Gliding, DFS), Walter Georgii, as best suited for this task on account of his experience and the respect he enjoyed in German and international science. Although Georgii had devoted himself to the improvement of 'his' research establishment, he could not resist this offer as he felt personally obliged to Baeumker. On the other hand, Georgii knew the power relationships in the RLM only too well. He feared that he would be exhausted by the internal power struggles. In order to defend himself against this, he made his agreement to take the job contingent on complete support from the scientific community, the RLM and especially its technical development departments.

What Baeumker and the scientific community demanded from the Nazi government was nothing less than the reversal of a decision which had been taken in October 1941 at the time of the reorganization of air armaments, namely the subordination of research to development. But the father of the reform of air armament production, Erhard Milch, agreed that organizational changes in the research sector were necessary. When the respected aircraft engineer Kurt Tank supported the idea of a reform of state control, Milch began to seek solutions. He found one in a compromise between subordinating research to development and allowing the scientists' self-determination in their own fields. Research would stay under the control of the Technical Office of the RLM. Apart from this he wanted to install a special research council for aeronautics. For chairman of the committee he wanted not the head of the DFS, like Baeumker, but the doyen of aeronautical research, Ludwig Prandtl. Friedrich Seewald, the head of the DVL in Berlin-Adlershof, was to serve as secretary of the council; Baeumker would be responsible for technical administration. Since Milch did not want to overrule the suggestion of the scientists he completed the four-man leadership team with Georgii.

Although Prandtl was completely overworked, he agreed. This reaction was typical of his deep-rooted notion of the duties of the citizen. To refuse an assignment from those in authority was beyond his imagination. He saw the advantage of the reform less in the four-man leadership team than in the committee that would serve under them, which was drawn from representatives from research and the aircraft industry. The chief purpose of the committee would be to use the stimulus coming out of the departments of the RLM to set up new research tasks and channel them onwards to the RLM (Research Leadership). This committee linking science and industry, however, was never established. Milch and the RLM preferred a bureaucratic solution. The Research Leadership would be composed of personnel from the former research department of the Ministry. The Research Leadership was left hanging, according to Prandtl's formulation, 'downwind' (cit. after Trischler 1992c: 248).

Did the experiment of allowing the scientific community to run its own affairs pay off for the Nazis? The answer is yes, at least partially. The members of the Research Leadership repaid the trust that Milch had placed in them by mobilizing their energies for the goals of the regime. Seewald and Georgii had in their careers always represented the position of outsiders, since research had had to gear itself to the needs of industry. Admittedly they had been also fighters for the freedom of science from political control. Now they went a big step further, indeed they changed fronts. From being champions of the autonomy of science they became supporters of political control. Seewald and Georgii called for a concentration of critical war work without concern for traditional research structures. The needs of the military should determine the direction that research should take. 'Our wish is that research fulfils its task in this critical stage of the war', said Georgii in 1944 (cit. in Trischler 1992b: 211).

From August 1942 the Research Leadership worked on a concept from which Prandtl and his colleagues expected a distinct increase in efficiency. In order to make possible the control of work in individual centres, research would be divided into individual disciplines led by an expert. These experts had the power to steer the work according to the instructions of the Research Leadership. Experts in neighbouring areas would be placed in groups and led by *Obleute* (seniors) whose main job was to coordinate cooperation between industry and the development branch of the RLM. While the internal control was handled by the experts, the seniors should ensure that development and testing ran smoothly.

The situation within the Research Leadership deteriorated towards the end of the war. Allied bombing made travel and communication much more difficult. In the end, the office of the Research Leadership had to be divided between north and south Germany. Nevertheless, the four-man body produced remarkable results. And here we find the important characteristics of big science in a nutshell again: project orientation, resource concentration, and the connection of the subsystems of state, science and industry:

1. Research work which was non-essential to the war effort was stopped. Resources were concentrated on projects which were important to the war. Information on what was vital to the war effort came to the Research Leadership in twice-weekly discussions with the *Generalluftzeugmeister*, in which research was represented by Seewald und Georgii.

2. Research plans were subject to a clear thematic focus. Especially good was the coordination of the work in the area of aerodynamics. In the central discipline of aviation research the influence of the elder statesman Ludwig Prandtl was critical. He ensured a high degree of internal communication with his former students despite the secrecy rules. Aerodynamics was by this stage the product of a reform which had taken place in 1943 independent of the other measures. A special committee on wind tunnels, composed of representatives from industry and research, was empowered to look at all tests and results. The committee determined that the tunnels should be divided equally between research and industry and that the results should be quickly made available where they were needed. These facilities, formerly completely blocked by orders from industry, were now freed up for research again.

3. The frequent complaints of the scientists that their work was being sabotaged by the restrictions on contacts with industry were finally taken into account. Even the control freaks of the RLM had to admit that it made no sense to have seniors and experts when at the same time the cooperation of the centres with firms was hampered by security restrictions. At the same time Georgii, who took over responsibility as head of the Research Leadership from Prandtl in 1943, succeeded in establishing contacts with the front through his personal relationships with leading Luftwaffe generals. Now at last the scientists got the chance to see the effects of their work on the war effort and take on board the wishes of the flying crews.

4. Through *ad hoc* special commissions pressing problems were processed in an unbureaucratic fashion. For research on high-speed flight a joint effort with industry was initiated in 1944 accompanying the fighter programme of the RLM.

Small Science or Big Science?

Even more than the Great War, the Second World War was a total war, which, through its length and intensity, managed to drag entire societal subsystems of the participating nations in its train. Wars like this challenge political systems to channel the resources of a society into war production, not only in the short term but also in the medium term. The First World War showed the importance of mobilizing science and technology. The technology of submarines and aircraft, developed on the eve of the war, expanded the action of the war to a third dimension. In the course of the military actions aviation had developed into a multi-purpose weapon system. Striving for tactical and strategic superiority became one of the most important goals of the conflict. Of all the warring nations it was Britain that most consistently answered the challenge. In November 1917, the Royal Air Force (RAF) was created as the third independent arm of the British military. One year later, right at the end of the war, the RAF had been ready to bomb Berlin with new four-engined Handley Page bombers, thus showing the shape of things to come (Edgerton 1991: 16–17).

The new military aim of gaining superiority in the air was formulated by the Italian general Giulio Douhet in 1921 in his most influential book, *Il dominio dell'aria*. When the book was translated into German in 1935, Hitler and Göring began to build up a powerful air force which would provide the military basis for victory in a transcontinental European war (Boog 1982; Murray 1986; Schabel 1994). During the pre-war years aviation research had a secure place in the Nazi rearmament programme. It was intended to provide the scientific basis for new weapon systems which would give the German Luftwaffe technical superiority in the medium and long term. The funds which were made available were correspondingly large. A traditional and conservative ministerial bureaucracy succeeded in giving research a high degree of scientific autonomy.

With the outbreak of war this period of autonomy and medium- and long-term planning came to an abrupt halt. The regime adopted tight controls to ensure that research was only directed at short-term objectives that would yield immediate rewards for the war effort. The military doctrine of the regime – the idea that enemies should be quickly subdued in *Blitzkriege* – dictated research philosophy. Scientific personnel were dispersed throughout industry and the battle front. The long-term goals of research projects were pushed into the background.

During the crisis of air armament in the middle of the war, the regime abandoned this approach and returned to the pre-war model that gave greater autonomy to the scientists. This change of course is remarkable. It would be a mistake, however, to see it as evidence of the Nazis' ability to react to the failure of the attempt to control science and thereby to credit the regime with any particular insight into the complexity of the innovation process. Rather it was a desperate attempt on the part of the technocratic elite, associated with the figures of Erhard Milch and Albert Speer, to couple research to armament development and production.

The utopian hope that science would conjure new weapon systems out of a hat gave new room for manoeuvre to the symbolic alliance between state technocracy and scientific self-mobilization in the last phase of the war. Like every symbiotic relationship, the alliance had advantages for both sides: under the leadership of Ludwig Prandtl and Walter Georgii research resources were concentrated on technology-oriented big science projects and thereby increased the efficiency of research in the armament sector. With the trump card of military necessity in their hands, the Research Leadership were actually able to recover a good portion of the scientists who had been drafted to the front (Ludwig 1974: 251–71).

Admittedly, control of the scientists' activities was only partial. The concentration on war-relevant projects really took place on the desks of the Research Leadership. In the chaos of the war years the chances were greater than ever for individual scientists to conduct their own work as 'small science'. How big these opportunities were can be demonstrated by the efforts of the Research Leadership to integrate the research capacities of the universities of Straßburg and Heidelberg into its war-related system. The Reich University of Straßburg was selected as a new centre of aviation research in 1941. The RLM, the university administration, the *Gauleiter* Robert Wagner and the Research Leadership managed to get themselves entangled in a pointless conflict. And for the scientists on the spot it was easy enough to play off one party against the other in order to pursue their own agenda. In Heidelberg the funds from the RLM were channelled into a long-shelved plan to create an institute for the study of world air transport. Once again the university succeeded in protecting its autonomy. Later in the war, in Heidelberg fields of such questionable value for the war effort as air law and language research expanded (Trischler 1992c: 271–4).

In conclusion: in the wake of the First World War the structure of big science was imposed upon aviation. The momentum created by the need for rearmament strengthened this trend. Big science required especially close cooperation between government, science and industry. The state as political actor was confronted with the challenge of coordinating this cooperation. The history of aviation research under the Nazis can be seen as a vain attempt to counter this trend. When the regime returned to the model of scientific self-administration in the middle of the war it was simply acknowledging the consequences of its own failure. It is difficult to judge whether the results of this self-administration were wholly positive. On the one hand, there was an increase in the concentration of resources on big projects. On the other hand, the scientists were given more room for manoeuvre, so that they were able to look beyond the dedicated research tasks to their own work. National Socialism and war offered room for big science and small science to work together.

A glance at national innovation systems in the post-war period shows that the coexistence of big science and small science in large technologies is by no means a rarity. Big science at the macro level of institutional large projects corresponds to small science at the micro level of scientific laboratory work. This coexistence hardly needs an explanation. What is decisive for big science is the way it uses small science within mission-oriented projects of the middle- and long-term type. The Nazis failed in their attempt to set the structural conditions for this development. And the chief obstacle was the polycratic structure of the Nazi dictatorship, which hindered communication between subsystems tending towards social and political self-determination.

From an international perspective the research system of the United States met this challenge much better than the Nazi regime. It was not only nuclear researchers who reacted to the challenge from Germany with the Manhattan Project. The National Advisory Committee for Aeronautics (NACA), which was answerable directly to the President, also countered the threat of German aeronautical superiority with new personnel and organizational reforms. In June 1940 Roosevelt appointed the NACA chairman, Vannevar Bush, as chairman of the National Defense Research Committee, which was shortly afterwards renamed into Office of Scientific Research and Development. The appointment provided the go-ahead for the mobilization of brains for war and accelerated the transfer of scientific research results to the business of aircraft production. For the nation's security, Bush declared, 'we must leave no stone unturned in research', and he added (cit. in Zachary 1997: 110):

> That there should be some direct relationship between military success and the *quality* of research in aeronautics is fairly obvious. Accurate information and a correct interpretation of research results are necessary. But *quantity* of research is also highly important. A large capacity to do research work is as essential as a large capacity to produce aircraft. It

would be foolish to create a bottle-neck for quality production because of lack of ability to obtain research results quickly. Under wartime pressures much may depend upon the ability of airplane designers to get correct answers to new problems quickly.

Adding it all together, it is safe to say that the course of history is largely being influenced by scientific investigations in aeronautical research laboratories. The outcome of the aerial battles of tomorrow are being decided today by the men who are working in wind tunnels, towing tanks, and engine laboratories.

Against this background the Manhattan Project loses its singular and spectacular aspect. Nuclear research and aeronautical research can be seen as the culmination of a metamorphic process, born out of a crisis of national security. In this way big science became a new form of institutionalized research.

References

Abelshauser, Werner (1984), 'The First Post Liberal Nation: Stages in the Development of Modern Corporatism in Germany', *European History Quarterly*, 14: 285–318.

Baeumker, Adolf (1944), *Zur Geschichte der deutschen Luftfahrtforschung – Ein Beitrag*, München: Eigenverlag.

Betz, Albert (1949), 'Ziele, Wege und konstruktive Auswertung der Strömungsforschung', *Zeitschrift des VDI*, 91: 253–8.

—— (1957), 'Entwicklungstendenzen der Forschung und ihre Gefahren', *Physikalische Blätter*, 13: 385–90.

Blenk, Hermann (1941), 'Die Luftfahrtforschungsanstalt Hermann Göring', in Karl Stuchtey and Walter Boje (eds), *Beiträge zur Geschichte der deutschen Luftfahrtwissenschaft und -technik*, vol. 1, Berlin: Reichsdruckerei, 463–51.

Boog, Horst (1982), *Die deutsche Luftwaffenführung 1935–1945. Führungs-probleme, Spitzengliederung, Generalstabsausbildung*, Stuttgart: dva.

Budraß, Lutz (1998), *Flugzeugindustrie und Luftrüstung in Deutschland 1918–1945*, Düsseldorf: Droste.

Cartellieri, Wolfgang (1963), 'Die Großforschung und der Staat. Gedanken über eine zweckmäßige Trägerform für öffentlich geförderte Großvorhaben der wissenschaftlichen Forschung und technischen Entwicklung', in Bundesminister für wissenschaftliche Forschung (ed.), *Die Projektwissenschaften*, Munich: Gersbach, 3–16.

Chadeau, Emmanuel (1985), 'Etat, industrie, nation: La formation des technologies aéronautiques en France (1900–1950)', *Histoire, économie et société*, 4: 275–99.

Constant, Edward (1980), *The Origins of the Turbojet Revolution*, Baltimore: Johns Hopkins University Press.

Edgerton, David (1991), *England and the Aeroplane. An Essay on a Militant and Technological Nation*, London: Macmillan.

Evans, Richard (1997), *In Defense of History*, London: Granta Books.

Friemert, Chup (1980), *Produktionsästhetik im Faschismus. Das Amt 'Schönheit der Arbeit'*, München: Damnitz.

Fritzsche, Peter (1992), *A Nation of Fliers. German Aviation and the Popular Imagination*, Cambridge/Mass. and London: Harvard University Press.

Galison, Peter and Hevly, Bruce (1992), 'The Many Faces of Big Science', in Peter Galison and Bruce Hevly (eds), *Big Science: The Growth of Large Scale Research*, Stanford/CA: Stanford University Press, 1–17.

Gundler, Bettina (1995), 'Das "Luftfahrtlehrzentrum": Luftfahrtlehre und Forschung an der TH Braunschweig im Dritten Reich', in Walter Kertz (ed.), *Technische Universität Carolo-Wilhelmina Braunschweig 1745–1995: Vom Collegium Carolinum zur Technischen Universität*, Hildesheim: Olms, 509–31.

Häfele, Wolf (1963), 'Neuartige Wege naturwissenschaftlich-technischer Entwicklung', in Bundesminister für wissenschaftliche Forschung (ed.), *Die Projektwissenschaften*, München: Gersbach, 17–38.

Hanle, Paul A. (1982), *Bringing Aerodynamics to America*, Cambridge/Mass. and London: MIT Press.

Hansen, James R. (1987), *Engineer in Charge: A History of the Langley Aeronautical Laboratory, 1917–1958*, Washington, D.C.: National Aeronautics and Space Administration, Scientific and Technical Information Office.

Heinzerling, Werner (1990), 'Windkanäle', in Ludwig Bölkow (ed.), *Ein Jahrhundert Flugzeuge. Geschichte und Technik des Fliegens*, Düsseldorf: VDI-Verlag, 304–33.

Homze, Edward (1976), *Arming the Luftwaffe: The Reich Air Ministry and the German Aircraft Industry, 1919–39*, Lincoln: University of Nebraska Press.

Kowarski, Lew (1949), 'Psychology and Structure of Large-Scale Physical Research', *Bulletin of the Atomic Scientists*, 5, Aug./Sept.: 186–204.

Lorenz, Chris (1998), 'Can History be True? Narrativism, Positivism and the "Metaphorical Turn"', *History and Theory*, 37: 309–29.

Ludwig, Karl-Heinz (1974), *Technik und Ingenieure im Dritten Reich*, Düsseldorf: Droste.

Macrakis, Kristie (1993), *Surviving the Swastika. Scientific Research in Nazi Germany*, New York and Oxford: Oxford University Press.

Marsch, Ulrich (1994), *Notgemeinschaft der Deutschen Wissenschaft. Gründung und frühe Geschichte 1920–1925*, Frankfurt a. M.: Lang 1994.

Mehrtens, Herbert (1996), 'Mathematics and War: Germany, 1900-1945', in Paul Forman and José M. Sánchez-Ron (eds), *National Military Establishments and the Advancement of Science and Technology. Studies in the 20ᵗʰ Century History*, Dordrecht, Boston and London: Kluwer, 87–134.

Murray, Williamson (1986), *The Luftwaffe 1933–1945: Strategy for Defeat*, London: Brassey's.

Nelson, Richard (ed.) (1993), *National Innovation Systems*, New York: Oxford University Press.

Price, Derek de Solla (1963), *Little Science, Big Science*, New York: Columbia University Press.

—— (1965), 'Is Technology Historically Independent of Science? A Study in Statistical Historiography', *Technology and Culture*, 6, 553–68.

Rabinbach, Anson G. (1976), 'The Aesthetics of Production in the Third Reich', *Journal of Contemporary History*, 11: 43–74.

Rip, Arie van (1992), 'Science and Technology as Dancing Partners', in Peter Kroes and Martijn Bakker (eds), *Technological Development and Science in the Industrial Age. New Perspectives on the Science–Technology Relationship*, Dordrecht, Boston and London: Kluwer, 231–70.

Ritter, Gerhard A. (1992), *Großforschung und Staat in Deutschland. Ein historischer Überblick*, München C. H. Beck.

Rotta, Julius C. (1990), *Die Aerodynamische Versuchsanstalt Göttingen – ein Werk Ludwig Prandtls. Eine Dokumentation ihrer Geschichte 1907–1925*, Göttingen: Vandenhoeck & Ruprecht.

Schabel, Ralf (1994), *Die Illusion der Wunderwaffen. Die Rolle der Düsenflugzeuge und Flugabwehrraketen in der Rüstungspolitik des Dritten Reiches*, München: Oldenbourg.

Simon, Leslie (1947), *German Research in World War II*, New York: Wiley.

Szöllösi-Janze, Margit and Trischler, Helmuth (1990), 'Entwicklungslinien der Großforschung in der Bundesrepublik Deutschland', in Margit Szöllösi-Janze and Helmuth Trischler (eds), *Großforschung in Deutschland*, Frankfurt a. M. and New York: Campus, 13–20.

Trischler, Helmuth (1992a), 'Aeronautical Research in the Third Reich. Organization, Management and Efficiency during Rearmament and War', in Horst Boog (ed.), *The Conduct of the Air War in the Second World War. An International Comparison*, New York and Oxford: Berg, 169-95.

—— (ed.) (1992b), *Dokumente zur Geschichte der Luft- und Raumfahrtforschung in Deutschland 1900–1970*, Köln: DLR-Mitteilungen.

—— (1992c), *Luft- und Raumfahrtforschung in Deutschland 1900–1970. Politische Geschichte einer Wissenschaft*, Frankfurt a. M. and New York: Campus.

—— (1994), 'Self-Mobilization or Resistance? Aeronautical Research and National Socialism', in Monika Renneberg and Mark Walker (eds), *Science, Technology and National Socialism*, Cambridge: Cambridge University Press, 72–87.

—— (1995), 'Großforschung und Großforschungseinrichtungen', in Peter Frieß and Peter Steiner (eds), *Forschung und Technik in Deutschland nach 1945*, München: Deutscher Kunstverlag, 112–23.

—— (1996), 'Die neue Räumlichkeit des Krieges: Wissenschaft und Technik im Ersten Weltkrieg', *Berichte zur Wissenschaftsgeschichte*, 19: 95–103.

Walker, Pery P. (1971), *Early Aviation at Farnborough: The History of the Royal Aircraft Establishment*, 2 vols, London.

Weinberg, Alvin M. (1967), *Reflections on Big Science*, Cambridge/Mass.: MIT Press.

Weiss, Burghard (1988), 'Großforschung. Genese und Funktion eines neuen Forschungstyps', in Hans Poser and Clemens Burrichter (eds), *Die geschichtliche Perspektive der Wissenschaftsforschung*, Berlin: TUB-Dokumentation, 149–75.

Witt, Peter-Christian (1990), 'Wissenschaftsfinanzierung zwischen Inflation und Deflation. Die Kaiser-Wilhelm-Gesellschaft 1918/19 bis 1934/35', in Rudolf Vierhaus and Bernhard vom Brocke (eds), *Forschung im Spannungsfeld von Politik und Gesellschaft. Geschichte und Struktur der Kaiser-Wilhelm-/Max-Planck-Gesellschaft*, Stuttgart: dva, 579–656.

Wüst, Walter (1982), *Sie zähmten den Sturm. 75 Jahre Aerodynamische Versuchsanstalt Göttingen*, Göttingen: DVFLR.

Zachary, Pascal G. (1997), *Endless Frontier: Vannevar Bush, Engineer of the American Century*, New York: Free Press.

Ziegler, Charles A. (1994), 'Weapons Development in Context. The Case of the World War I Balloon Bomber', *Technology and Culture*, 35: 750–67.

LUITGARD MARSCHALL

Consequences of the Politics of Autarky: The Case of Biotechnology

Introduction

National Socialism permeated large areas of the cultural, political and economic sphere, and science and technology were by no means exempt from this. During the Third Reich they were massively functionalized in order to serve the political, military and ideological aims of the Nazis. More than ever before, the development of technology had to be adapted to the prevalent national conditions and requirements. The autarky of Germany was to be brought about as rapidly as possible, making use of all the technical means available. Outstanding among these were the production methods of the chemical-pharmaceutical industry, which will be the main focus of the following discussion.[1]

Then as now, in this industrial branch there were basically two different paths open for production. Numerous substances can be produced not only chemically but also in a biotechnological way by means of micro-organisms like bacteria, yeasts or moulds. Ordinary alcohol (i.e. ethanol) for example, can be a product of chemical synthesis as well as of fermentation. Because of their mutual substitutability the two production methods competed with each other in several fields. But they also differ in several aspects and are therefore able to serve quite different industrial, political and economic needs. With respect to the raw material base, chemical synthesis traditionally used coal as primary raw material, which – through the influence of catalysts, high temperature and high pressure – was converted into the desired product. Later on, coal was

111

replaced by other fossil raw materials like petroleum or natural gas.[2] In contrast, biotechnological production relied almost wholly on agricultural raw materials like potato and grain starch or sugar. These carbohydrates are transformed through micro-organisms into diverse products including alcohol, glycerin or penicillin. Owing to their theoretical foundation in organic chemistry, the research and production methods of chemical synthesis were regarded already at the beginning of the twentieth century as extremely science-based and therefore as modern and innovative.[3] Biotechnology, in contrast, was for a long time considered to be an empirical and backward production technique, just because of its lack of theory and science-based methods (Buchholz 1979: 91; Marschall 2000: 185–202).

Owing to massive state interventions during the Third Reich the coal-based high pressure synthesis managed to make its way as the predominant production technology. It impressed by its state-of-the-art solution methods, but was – at least in some areas – so costly and energy-intensive that it could only be implemented and maintained by government subsidies. On the one hand, this created a 'high technology'.[4] On the other hand, the concentration on coal-based chemical synthesis led at the same time to the neglect of alternative technologies, among them petroleum-based chemistry and biotechnology. The consequences of this became evident only decades later in the serious lag that Germany exhibited in biotechnological production.

In this chapter I shall outline how the steering manoeuvres of the Nazis influenced the development of biotechnology both directly and indirectly. To state the conclusion in advance: between 1933 and 1945 biotechnological processes were used only to compensate the deficits of chemical syntheses. Biotechnology was thus relegated to special processing niches. However, this relegation of biotechnological processes in favour of chemical ones had already started prior to 1933 and, after 1945, it culminated in further retardation. This of course indicates a continuous process. Nevertheless, the development in the Third Reich demonstrated particular features which had a lasting influence on the further course of biotechnology. Therefore the central aim of my essay is to elucidate these special aspects and their long-term consequences. Comprehension, however, presupposes knowledge of biotechnological development before, during and after the Third Reich. So I shall endeavour to survey this, starting at the turn of the century and continuing until the 1970s. As the study will mainly focus on the years between 1933 and 1945, the picture of the development in the preceding and following years must remain quite sketchy.

I shall start by conveying some basic ideas which constitute the theoretical structure of my investigation. That the development of technology is a result of different social processes is meanwhile commonly accepted in the history of technology (Rammert 1992). Social requirements, economic calculations and pressures, technological criteria of efficiency and politically imposed regulations have a deep impact on the formation of technology. Simultaneously technological development also shows a certain robustness in relation to those social influences. Among other things, this is due to the material character of the technical artefacts. Tools and machines are generally conceived for specific application purposes and are thus not arbitrarily usable. An atomic power station simply cannot be used for the generation of solar energy. Likewise a chemical high pressure reactor is not suitable as a fermenter for microbiological conversions. Objective limitations like these cause technological development to be subject to its own dynamics and makes its external control increasingly difficult, if not impossible (Weingart 1989: 9).

This double-faced character of technology – responsiveness to social pressure and independent dynamic development – can be grasped through the metaphor of so-called technological trajectories. The trajectory model comes from the innovation economy, where it serves as an explanatory model for the technological change in individual industrial enterprises and branches (Nelson/Winter 1977: 36–76; Dosi 1982: 147–62).[5] It is based on the observation that technological development is more strongly determined than had hitherto been assumed. In order to keep their market position, industrial agents have to take long-term decisions even under conditions of great uncertainty with respect to the potentials of new technologies. As a consequence their future research and development potentials are tied to technological paths. In the course of time such a path tends to gain its own dynamic and becomes more and more difficult to reverse. This path dependency can lead to an exclusion of alternative methods. Thus the decision in favour of a certain production technology often implies a simultaneous decision against technological alternatives.

The metaphor of technological trajectories provides a viable explanatory model for the development of industrial biotechnology in Germany. As is well known, the German chemical industry decided early on to follow the path of chemical synthesis (Landau/Rosenberg 1992). This decision proved to be extraordinarily successful and contributed to its gaining a leading position in synthetic organic chemistry. It had some opportunity costs, however: up to now the fact has been ignored that the early commitment to chemical synthesis and its intensive expansion was detrimental to the development of biotechnology. For a long period in the twentieth century biotechnology played the part of a 'loser technology'. Its academic and industrial development was neglected in Germany until the 1970s in favour of chemical synthesis.[6]

The interaction between the two production technologies requires that in writing the history of biotechnology one must also take into account the development of chemical synthesis. In the subsequent survey I shall show that the shaping of the technological path of chemical synthesis – and hence the stagnation in the development of biotechnology – happened gradually in several stages. These individual stages were determined by diverse factors, which will be outlined in the following pages. The period of the Third Reich is of utmost importance because during these years the path of chemosynthesis was stabilized through massive state intervention to such an extent that it could develop a dynamic of its own in the post-war era. For biotechnology this led to a permanent relegation to a very few processing niches.

Biotechnology as an Alternative Production Method

The years from the turn of the century to the First World War constitute the heyday of biotechnology. In this early stage many new applications were developed (Bud 1993: 27–45; Marschall 2000: 89–105). Before that time microbiological production was almost exclusively used in the traditional fermentation industries. Now it found its way into industrial fields as diverse as sewage treatment, the leather industry or the chemical-pharmaceutical industry. In the latter branch, chemical and biotechnological methods already co-existed in production. While organic dyestuffs and drugs were preferably synthesized chemically, various commodity chemicals like lactic, butyric or acetic acid, aceton or diverse alcohols were produced by means of micro-organisms. The biotechnological products of this initial stage were mainly compounds with a simple chemical structure, which were produced on a large scale. Thus, in stark contrast to its use today, biotechnology was for the most part used as a means for mass production of cheap solvents and basic chemicals.[7] Not only in industrial production, but also in the field of research and development many problems were tackled by using biotechnological as well

Figure 8   Advertisement for alcohol-driven engines and machines, 1902.

as chemical approaches. Already in 1897, Emil Hansen, director of the physiological department of the Carlsberg Laboratory in Copenhagen, judged that whereas in former times chemistry had dominated the realm of problem solving, now biology had gained a place by its side (Hansen 1897: iv). Grave problems like the feared shortage of natural nitrogen fertilizer, petroleum or rubber seemed solvable in the near future by both methods of production (Marschall 2000: 89–105).

However, owing to the lack of technical maturity, the prospects of both methods were still difficult to assess. The arguments in favour of biotechnology were that its technical procedures were simple and the agricultural raw materials cheap. Indeed, at that time the raw material supply for biotechnology was secured through the yields from prospering national agriculture, mainly from potato cultivation in the eastern parts of Germany. It was supplemented by grain, imported in large amounts from overseas.[8] Thus, in this first stage of development, one cannot find serious economic or political factors counting against biotechnology. As Max Delbrück, director of the Institut für Gärungsgewerbe und Stärkefabrikation zu Berlin, one of the largest research centres of the German fermentation industry, said, it seemed rather a good idea for the chemical industry sometimes to leave the retort behind and instead to intensify the use of bacteriological methods (Delbrück 1902: 699).

Against this background it is interesting to observe that different branches of the heterogeneous chemical-pharmaceutical industry showed distinct reaction patterns towards biotechnology. On the one hand, the small- and medium-sized firms of the traditional pharmaceutical sector were quite engaged in biotechnological production, while on the other hand the large companies of the chemical dyestuff industry displayed almost no interest. As a result, there was a quick diffusion of biotechnogical methods into the pharmaceutical, but none into the dyestuff industry. This specific behaviour of the two branches is evidence for their different cultures of entrepreneurship and therefore a consequence of their different historical roots.[9] Although nowadays the classical division of the chemical-pharmaceutical industry into two separate parts doesn't make sense any more, the differences loomed large until the middle of the twentieth century. They manifested themselves in several ways: both branches had different technical traditions and product lines and varied in the standards of education among their employees and therefore in their styles of research and production.

To put it in a nutshell: almost from its beginnings, the dyestuff industry was regarded as the prototype of a science-based and innovation-oriented industry (Radkau 1989: 41) In its research laboratories chemists made systematic use of the abstract scientific knowledge gleaned from universities in order to produce new dyes and drugs. Research and production were based on the theories of structural chemistry.[10] This close linkage of theory-guided research and industrial production guaranteed continuous innovation. At the same time it contributed both to the scientist's self-image and the ideals of the chemists in this branch. Thus it had a deep impact on their perception and their selection of new technologies. From the point of view of an academically educated chemist the theoretically ill-understood and therefore mainly empirical procedures, which were typical of biotechnological research and production, must have looked pre-scientific and therefore anachronistic.[11] But it was exactly this disdainful attitude which finally led to that underestimation of biotechnology that would characterize its later stages of development.

The science-based dyestuff industry must be contrasted with the traditional pharmaceutical industry. The latter's enterprises often grew out of pharmacies and were run by more technically trained pharmacists.[12] Typically, these enterprises relied on the use of traditional empirical methods to manufacture medicaments, e.g. the extraction of drugs from plants or animal organs. To do this success-fully, one needed practical skills and personal experience, but not necessarily detailed knowledge of the theoretical foundations of the processes involved. Biotechnology fitted well into that artisanal tradition: its empirical methods did not presuppose a deep theoretical understanding. They rather depended on practical knowledge in dealing with micro-organisms and on technical means for cultivating them. Another reason why the mainly small-sized pharmaceutical firms with no strong financial background entered biotechnology was the fact that biotechnological processes were technically uncom-plicated and therefore not too expensive. Not surprisingly, the number of biotechnologial producers of lactic acid grew from zero to twenty-five between 1895 and 1907 (Schäfer 1907: 177).

Shaping the Path of Chemical Synthesis

The diffusion of biotechnological processes continued during the First World War, when strategically relevant methods of fermentation like the production of glycerine or fodder yeast were newly introduced. After the war, however, it came to a sudden end: between 1918 and 1933 a large number of the processes that had been previously introduced were given up again. From that time on, biotechnology would be increasingly relegated to a few technical niches. The relegation happened in several steps. In the second stage of the development – the years of the Weimar Republic – it was triggered by economic factors which were related to a dramatic change in the raw material supply. First of all agricultural production during the First World War suffered an almost complete breakdown. It would take years to secure the nutrition of the population again. Even more decisive was the loss of important agricultural production sites due to the reorganization of frontiers in the Versailles Treaty, especially those for the cultivation of potatoes in the east. Now that agricultural products had become scarce and expensive, biotechnology was robbed of its raw material base.[13] Before the war, the abundance of cheap agricultural starting materials was an important incentive for the biotechnological mass production of commodity chemicals. After the war, these biotechnological methods could no longer compete with chemical processes: the lack of cheap raw material undermined the position of biotechnology (Marschall 2000: 105-15). As a consequence many promising research programmes of the pre-war area were cancelled; no comparable new biotechnological projects got started. For example, the pre-war idea of substituting fermentation alcohol for gasoline was given up in favour of the chemical solution, the coal-based high-pressure synthesis of fuel. So biotechnological research and development activities were given up, while at the same time the corresponding chemical efforts were intensified – in complete contrast to developments in the United Kingdom and the United States, where the favourable raw material endowment contributed to the furthering of biotechnology (Bud 1988: 442; Bud 1993).

In those pharmaceutical enterprises already using biotechnology, the precarious resource situation caused a decline in profitability and lean production years.[14] Not only did the number of firms relying on biotechnology diminish, the general applications for biotechnological processes also began to change. Previously biotechnology was intended as a method for the mass production of commodity chemicals. From now on, it was increasingly used for the production of special sub-stances with a chemically complex structure, for example enzymes or, in later years, antibiotics. Some of these compounds could not be produced at all by chemical methods, while the production of others would have been extremely expensive. So biotechnology could hold its own in these fields of production.

While the application of biotechnology in the pharmaceutical industry was decreasing, the chemical dyestuff industry saw the triumph of coal-based high pressure synthesis. After the introduction of the Haber-Bosch process during the war, this new technological line was broadly extended in the post-war era. In BASF, Carl Bosch argued for its further development by pointing to the high expenditure already devoted to the Haber-Bosch process. Moreover the under-utilized production plants constructed during the war to carry out ammonia synthesis could thus be kept in production. Furthermore, the firm would be able to rely on its own expert and skilled personnel doing the research.[15] Hitherto only cyclical compounds like benzol had been extracted from coal. Now the aim was to produce even aliphatic (noncyclical) compounds out of coal. During the Weimar Republic BASF succeeded in synthesizing key aliphatic substances like methanol, ethanol, aceton, acetic acid and also synthetic fuel by means of high-pressure synthesis (Welsch 1981: 151). Thus biotechnology lost its formerly most promising fields of application. To summarize the second stage: during the Weimar Republic the shortage of agricultural raw materials blocked, on the one hand, the further development of biotechnology in research and application. On the other hand, the large-scale enterprises of the synthetic dyestuff industry, which had in the meantime merged to form the influential market leader I.G. Farben AG, took decisive steps on the path of coal-based high-pressure synthesis.

Governmental Path Stabilization: The Subsidiary Function of Biotechnology

Although the path of chemical synthesis was paved before 1933 by industrial decisions, its future success strongly depended on government support. Thus the third stage of development, comprising the years of the Third Reich, can reasonably be characterized as a period of governmental path stabilization. In 1933 a policy of active national autarky was started under the Nazis. As already mentioned, their influence did not lead to a change of the direction of technological development. On the contrary, chemical high-pressure synthesis was now used even more intensively. The primary aim of the economic policy of the National Socialists was to overcome the existing dependency on raw materials from abroad: apart from coal, Germany had to import almost all strategically relevant materials, among them petroleum, lubricants, rubber, natural fibres, fats, proteins, carbohydrates, iron ores or light metals (Plumpe 1990: 546–90). Yet the national raw material limitations seemed to be surmountable, at least in the long run, through chemical high-pressure synthesis. All hitherto imported raw materials were to be substituted in future by chemical surrogates, synthezised from domestic coal. As Claus Ungewitter, the head of the Wirtschaftsgruppe Chemische Industrie (Economic Group for the Chemical Industry), put it: the chemical synthesis of natural substances and surrogates revealed new and unconventional possibilities for Germany to make available what nature was not able to deliver in sufficient quantity or quality (Ungewitter 1938a: 7). According to Carl Bosch, director of the board of I.G. Farben, these possibilities were unlimited: in future there would be no natural substance that could not be produced synthetically from coal. Even the sphere of human nutrition would be revolutionized by synthetical chemistry (Radkau 1989: 269).

Nevertheless, such a surrogate policy wasn't something new for the chemical industry. The dyestuff enterprises had pursued it for commercial reasons ever since they had come into existence. Taking a look back in 1930, Carl Bosch categorically declared in conversation with the vice president of DuPont: 'We suppressed natural madder, then natural indigo, now we are suppressing Chile nitrate, also the wood distillation industry and we will soon supplant petroleum by hydrogenated coal' (Marsch 2000: 82). But what was new after 1933 was the decision that the production of surrogates was to be realized 'without any consideration of costs' in order to achieve economic self-sufficiency and military rearmament.[16] Owing to the extensive interventions of the National Socialists, the methods of coal-processing chemistry were further exploited and expanded into strategically important spheres, by simultaneously ignoring economic calculations and technological criteria of efficiency.

In core projects like the synthesis of fuel and rubber the interests of I.G. Farben and the Nazi government seemed to overlap, at least at the beginning. For both projects the research work had been started long before the Nazi regime. In 1933, during the nadir of the global economic crisis, when their further expansion no longer seemed viable from a market-economic point of view, the Nazi government created the necessary economic incentives for subsequent implementation (Plumpe 1990: 591). Ultimately it was because of public measures like targeted subsidization of research and development, investments in production plants, price and sales guarantees for surrogates and preferential allocation of scarce raw materials that the path of high-pressure synthesis during the Third Reich was continued and stabilized.

Economic principles in fact played only a minor role. This is evinced especially clearly by the examples of fuel and rubber synthesis. As already mentioned, both projects were initiated long before the Nazis came to power. In the case of synthetic fuel in 1924 the first test plant had been put into operation in Oppau. By 1927 so-called Leuna petrol had been launched (Welsch 1981: 118). However, as a consequence of the global economic crisis, the price for petroleum had slumped so drastically that the chemical synthesis of fuel was no longer viable from an economic point of view. The price of one litre of imported petroleum fuel was 5 Pfennig in 1933, while the production costs for synthetic fuel amounted to 19 Pfennig per litre (Welsch 1981: 119). Because the product could not compete internationally, the I.G. Farben group considered the closure of its hydrogenization plants. This was only prevented by the so-called petrol agreement of December 1933, which vouchsafed a national price and sales guarantee for synthetic fuel (Plumpe 1990: 271). Until 1936, the production of this surrogate already amounted to 42 per cent of the total fuel production in Germany. In the same year the synthetic fuel project was integrated into the national armament programme, where, until 1941, it used up almost half of the investment funds of the Four-Year Plan (Braun 1992: 34).

The synthetic rubber project had also reached its cost limits because of the price slumps for natural rubber as a result of the global economic crisis. In 1932, I.G. Farben discontinued its already far advanced research work. Owing to public investment it was resumed in 1933 and subsequently continued on a far larger scale. In 1927 I.G. Farben spent 0.1 million Reichsmarks on research and development costs for rubber synthesis. As a result of the world economic depression the investment increased at the end of 1929 to only 3 million Reichsmarks. After the standstill of the work in 1932 and the negotiations with the Nazis, in 1933 I.G. Farben finally invested at one blow 8 million Reichsmarks (Welsch 1981: 155). Henceforth, rubber synthesis was one of the core projects of the National Socialist autarky programme.

As both examples show, the implementation of these large-scale projects was primarily due to public intervention. Political-ideological reasons were decisive for the choice of technology - economic and technical criteria being only of secondary importance for the National Socialists. From an economic point of view, world economic depression had caused the prices of petroleum and natural rubber to fall to such a degree that importing them seemed to be preferable to their far more expensive synthetic production. As far as technical efficiency was concerned, the high-pressure procedures with their enormous expenditure of energy seemed inferior to the biotechnological production techniques.[17]

Abroad, Germany's commitment to coal as a primary raw material, and therefore the production of surrogates, was ridiculed. In 1937 a French author recommended that the Germans create a 'surrogate stomach' in order to digest all the 'surrogate products' (Ungewitter 1938a: 125). In Great Britain the influential spokesman of the British chemical community, William J. Pope, had already warned in the 1920s against imitating the German path of replacing natural substances. According to him, the chemical processes, which had been dependent on brute-force, high-energy reactions, were no longer elegant. In the future, it would be even more intelligent to emulate the gentle conditions of nature by using nature's own reactors. From Pope's point of view, Germany, without imperial possessions and therefore with few raw materials, was virtually compelled to make use of coal as a starting material to supply both energy and surrogates. In contrast, the British Empire would be able to use sunlight and the power of plants to produce an almost limitless supply of agricultural raw materials, which provided the basis for biotechnological production (Bud 1993: 46).

It is remarkable that Pope held virtually the same opinion as the German university chemist and I.G. Farben board member, Albrecht Schmidt. In 1934 Schmidt, too, considered biotechnology to be an extremely efficient production technology, which would turn out to be, in the long run, even superior to chemical synthesis (Schmidt 1934: 386, 614). However, its sensible and successful application would depend primarily on the geographic location of the user country. In the southern countries, owing to the favourable climatic conditions, there would be longer vegetation periods, which enabled far higher agricultural yields. The 'sunny countries' thus would have a sufficient amount of renewable raw materials for the biotechnological path. Germany, in contrast, was, in the words of Schmidt, a climatically disadvantaged 'non-sunny country' (Schmidt 1934: 386). Its main domestic resource was coal, while agricultural production had to be supplemented by imports from abroad. Therefore Germany was predestined for the application of coal-based chemistry. In the eyes of Schmidt and other Nazis, the domestic raw material endowment thus constituted the main argument in favour of chemical high-pressure synthesis and against biotechnology. Hence the encouragement of coal-based chemistry by the National Socialists was purely a political decision and not a rational choice between alternative production technologies.

Accordingly, during the Third Reich the development of technology was adapted to the domestic raw material supply. Since Germany was well-endowed with coal, the use of coal-based chemistry furthered the economic-political and military goals of the National Socialists.

By contrast, national agricultural production and the nutrition industry were far from being self-sufficient; Germany had to import a considerable quantity of agricultural raw materials (Petzina 1968: 91–6). The application of biotechnology on a large scale would have intensified this dependency and therefore contravened the autarky plan of the Nazis. However, at that time, one could not totally dispense with fermentation processes. As a consequence, biotechnology was put to a subsidiary use. Its application spectrum was limited to those spheres where chemical synthesis reached its technical and economic limits. In other words, biotechnological processes could only be used to compensate for the weaknesses and deficits of chemical synthesis. Thus, the volume of biotechnological production declined continually between 1933 and 1939. With the outbreak of the war, a renaissance of long-since discontinued biotechnical processes took place, which dated from the time of the First World War and served to produce vital war products, for example glycerine or fodder yeast. Notable technological breakthroughs and innovations were not achieved in Germany, however.[18] The limited application of biotechnology was rather based on minimal research efforts which were applied to extremely target-oriented individual projects but not to systematic research work. Altogether, the technical level of biotechnology during the Third Reich changed only imperceptibly. Its main characteristic was rather the relegation of its processes to a very few technical niches, which can be attached to three main application spheres.

First, it was a matter of the effective use of organic waste materials, above all, waste wood. According to Claus Ungewitter from the Wirtschaftsgruppe Chemische Industrie, a utilization of waste was indispensable because of the growing 'hunger for raw materials' in Germany. Decisive for these activities would be the fact that the national economy had primacy in German economic policy (Ungewitter 1938b: 164).[19] In his eyes the first and most important step, of course, to achieve independence from foreign raw materials had already been taken by establishing the coal basis of the chemical industry. But this would be only the beginning: in future, supplementary to coal, other domestic raw materials and waste products had to be opened up and processed. As he put it in his programmatic essay *The Processing of the Worthless*, the chemical industry would in future use new organic sources of raw materials, which would otherwise be useless for nutrition or animal feed (Ungewitter 1938a: 111):

For example the industry will find applications for mushrooms, certain reeds, wild fruits such as horse chestnuts, reeds, algae, seaweed, peat. These materials will be used to produce oils and fats, alcohols, sugars, starch and proteins, pharmaceuticals, especially biochemical catalysts like vitamins and hormons and alkaloids, further artificial fabrics and plastics, dyes, waxes, styrene, oils, solvents, and countless other organic preparations.

Hitherto unused organic materials were therefore to be converted into high-quality products. Yet such refinement processes could not be effected with the methods of synthetic chemistry. At best biotechnological processes were suitable for this, albeit to a limited extent. The so-called *Holzverzuckerung*, which means the production of wood sugar and its conversion into alcohol or fodder yeast, achieved actual importance in the recycling of waste (Ungewitter 1938b: 107–11). In this combined process, consisting of a chemical and a biotechnological step, waste wood was initially split into glucose through the addition of mineral acids, and subsequently, by means of yeast, metamorphosed into alcohol. The resulting glucose could also be used as a raw material for cultivating special yeasts, which were rich in proteins and therefore suitable as fodder. Thus, in 1943, for example, the wood distillation plant at Holzminden produced 360,000 litres of alcohol per month which served as fuel additive (Dellweg 1982: 19). Another important process was the recycling of sulphate waste water, which was generated in large quantities during the manufacture of cellulose staple fibre. It still contained between 2 and 3 per cent fermentable carbohydrates, which, by means of yeasts, could be converted into alcohol or fodder yeast. So in 1938, 700,000 hectolitres of so-called *Sulfitspiritus* were gained by this means and exploited as fuel additive. This was one quarter of the amount needed at that time (Ungewitter 1938b: 119).

The second domain of biotechnology during the Third Reich was the sphere of nutrition. As mentioned previously, the situation, as far as an independent foodstuff supply was concerned, was especially poor. With regard to the agricultural fat and protein production, autarky was out of question. In the early 1930s a 'protein gap' of 30 per cent and a 'fat gap' of as much as 40 to 50 per cent was ascertained (Petzina 1968: 95). Here, too, the gaps could not be overcome by the methods of synthetic chemistry. Due to their complex stereoscopic structure, proteins were not yet synthesizable, while the chemical production of fats at best led to inferior quality, unsuitable for human consumption (Ungewitter 1938a: 101). In contrast, even before the First World War it had been possible to create high-quality proteins and later also fats from carbohydrates by means of biotechnological processes. These well-known biotechnological processes were revived in 1939 with the building of several new production plants for the cultivation of protein yeast (Dellweg 1974: 31).[20]

A third, indispensable niche for biotechnology was the production of drugs of a complex chemical structure, including those with a specific stereoscopic structure. In this special sphere biotechnological processes are, on principle, superior to chemosynthetic ones. The industrial production of certain substances like vitamin B_2 or C could be brought about, for example, by means of micro-organisms or by a combination of chemical and microbiological processes. As Claus Ungewitter put it, biotechnology constituted a 'methodic aid' for chemosynthesis in the case of vitamin C (Ungewitter 1938a: 104). The synthetic production of this vitamin consisted of several chemical steps and only one microbiological step. Yet the microbiological assistance was indispensable for the profitable industrial production of vitamin C on the basis of glucose as a cheap starting material.[21]

All three areas demonstrate the subsidiary function of biotechnology in specific technological niches. The confining of biotechnology within these niches was steered by the raw material market or by the public raw material regulation. Thus the regimentation of the 1936 Four Year Plan caused a considerable decrease in biotechnological production, by additionally reducing the import of agricultural raw materials. The resultant shortage of agricultural products was to be counteracted, in accordance with governmental stipulations, through their far more intensive usage in all spheres. Furthermore, certain agricultural raw materials were given new applications. Potato starch, hitherto primarily used as fodder and raw material for biotechnological production, was now to serve primarily for human nutrition. Hence the percentage of available raw materials in biotechnological production decreased, causing a considerable price increase (Dehio 1966: 39).

In general, the government measures during the Third Reich caused a one-sided enhancement of the methods of coal-based chemical high pressure synthesis. Its application sphere expanded - coal-processing chemistry soon became the dominant production technology in I.G. Farben. In the pharmaceutical firms, the use of biotechnology was meanwhile reduced again, apart from its temporary flourishing in singular operational niches. As there were no remarkable innovations with respect to products and processes, the scientific and technological development of biotechnology kept on stagnating. This was also due to the fact that many organic chemists still held the firm conviction that chemical synthesis and its products were superior to biotechnological production. This attitude shows itself clearly in the belated and half-hearted engagement of Germany's big industry in penicillin research. This natural antibiotic is the metabolic product of a certain mould fungus. It is obtained by biotechnological means. In Great Britain and the United States the search for an industrial process for the biotechnological production of penicillin immediately started after the first report of its antibiotic effects in the scientific literature in August 1940. German chemists, however, clung to the synthesis of drugs; the importance of penicillin was underestimated. They believed that with synthetic sulfonamides they had the better drugs. Only when the therapeutic superiority of penicillin could no longer be ignored did German industry start doing research, but on a comparatively small scale. As a consequence of this and of inefficient public subsidizing, Germany was not able to develop an efficient procedure for the production of penicillin until 1945 (Pieroth 1992: 68–100).

The governmental definition of biotechnology as an ancillary technology had important consequences for the post-war era. Because of its almost exclusive use in domains where chemical synthesis had its deficiencies, biotechnology was later on commonly considered to be a mere tool for assisting chemistry. In the long run this perception consolidated its status as a subsidiary technology in the shadow of chemical synthesis.

With regard to developments before 1933, the interventions of the National Socialists did not lead to a break in technological development. On the contrary, they consolidated and accelerated the ongoing shaping of the trajectory of high pressure synthesis, on which the dyestuff industry had already decided during the Weimar Republic. Yet only the cooperation of I.G. Farben with the Nazis rendered possible such an expensive high technology, which indeed fitted the autarky plans, but was economically highly inefficient. By the end of Nazi rule the path of chemical high pressure synthesis was sufficiently stabilized that it subsequently displayed a dynamics of its own, even after the demise of the political and economic conditions that had supported it previously.

Biotechnology as a Niche Technology

Thus the fourth stage of the development covers the period between 1945 and the late 1960s. In what follows, my investigation will be concerned only with the industry in the Federal Republic of Germany. One can observe a further path extension, but now involving the pharmaceutical enterprises, too. Again, the economic and political environment had changed. To enter the international market anew, the successor companies of I.G. Farben, Hoechst, Bayer and BASF, changed their raw material basis from coal to petroleum in the 1950s (Stokes 1994). With the step towards petroleum-based chemistry, they did not need to abandon high pressure synthesis. On the contrary, their important strategies for growth and innovation were now as before based on the methods of organic chemical synthesis (Buchholz 1979: 70). The further expansion of chemical synthesis is demonstrated by the plethora of new synthetic compounds that have since flooded the market. It does not count against this diagnosis that chemical giants like Hoechst and Bayer in the late 1940s and early 1950s started with the biotechnological production of penicillin and other antibiotics. Their entry into biotechnology was based solely on US licences (Pieroth 1992: 107–10). No internal biotechnological research laboratories were set up in these enterprises at that time. Moreover, the biotechnological manufacture of antibiotics did not constitute more than a fringe segment of their total activity.

A new aspect emerging in the post-war era was that now the traditional pharmaceutical firms, too, plunged into chemical synthesis. Hitherto this technology had primarily made its way in the enterprises that stemmed from the erstwhile synthetic dyestuff industry and later merged into I.G. Farben. But now chemical synthesis became very important in the firms of the pharmaceutical industry, too, where it soon became the predominant production technology. After 1945, pharmaceutical enterprises that had hitherto been the mainstays of biotechnological production moved farther and farther away from their traditional techniques like extraction or biotechnology and turned towards chemical synthesis. The former pioneers in biotechnology, enterprises like Boehringer Ingelheim, Merck and Röhm, thus neglected their experience and knowledge about biotechnological production, which had been accumulated for decades.[22] Their retreat from biotechnology was due to several factors. First of all, with the change to petrochemistry and therefore with the introduction of the easily available and cheap raw material, petroleum, the competition

Figure 9   Production of penicillin at Hoechst (c.1960).

between chemical and biotechnological methods had been intensified. Secondly, the pharmaceutical industry, too, had in the meantime turned gradually from a technology-based into a science-based industry. This is indicated by the further differentiation of research structures in the enterprises, the higher educational standard of the personnel, and the increasing significance of theory-governed research and production methods, mainly deriving from organic chemistry. The result was a close adjustment in technological development between the chemical and pharmaceutical industries.

But this was not the environment within which biotechnology, with its low level of theoretical sophistication and empirical methods, could flourish. In contrast to organic synthesis, its potential had not been advanced during the last decades. Now, as before, chemists mainly judged biotechnology to be theoretically underdeveloped, representing a complex field with an inadequate theoretical structure (Buchholz 1979: 93, 101). Like the developments in the dyestuff industry at the beginning of the century, biotechnological procedures now hardly awoke interest in the pharmaceutical industry. Here, too, the growing success of chemical synthesis led to a change in perception, to an underestimation of biotechnology. The reason was that in the pharmaceutical enterprises the key positions meanwhile were held by scientifically trained and science-oriented chemists. Confronted with a choice between chemical and biotechnological methods, they regularly decided – in accordance with their knowledge and experience – in favour of chemical synthesis.[23]

In retrospect, the withdrawal from biotechnology was accompanied by quite a number of erroneous decisions, due to an altered perception and evaluation of biotechnology. After its long use as a subsidiary technology, in the post-war era many chemists thought of it in terms of a mere 'tool of chemistry'. This derogatory attitude and their scientific background and research tradition made them blind to the innovative potential of biotechnology. Thus after the Second World War, a further reduction in biotechnological methods took place, while pharmaceutical firms invested almost only in chemical synthesis.[24] To illustrate this with a concrete example, let us consider Röhm, a company in Darmstadt whose main products had previously been enzymes. After 1945 Röhm did not develop biotechnology any further. Instead it expanded one-sidedly into the profitable production of plastics (e.g. the safety glass *Plexiglas*). Even when in the 1960s the demand for enzymes in the food and detergent sector was steeply rising, Röhm continued with this strategy (Marschall 1997: 310). The Danish firm Novo (today Novo Nordisk) then entered the new market for enzymes. The advantage Novo gained by its early entry into this market has not been made up by any German enterprise up to this day.

Despite the free market conditions of the post-war period and regardless of the increasing demand for certain biotechnological products, in West Germany biotechnology remained restricted to a few technical niches. In contrast, the expansion of chemical synthesis was effected on a broad front in the chemical as well as in the pharmaceutical industries.

Counter-measures

The situation remained unchanged until the early 1970s. Then a new turning point emerged. In this last stage of the development, government and industry in the German Federal Republic tried to steer a new technological course (Jasanoff 1985; Giesecke 1998: 227–305). Again the background situation had changed and had led to a new perception of biotechnology. The economic pressure of two oil crises, a growing public sensitivity towards ecological matters and, last, but not least, decreasing turnovers for standard chemical products, forced the industry to look for alternative production methods and products. This search led to a re-evaluation of biotechnology.[25] Once its agricultural raw material base had been an argument against its extended use; now it was considered to be an essential advantage, given the scarcity of fossil resources. Previously its methods were judged to be empirical and therefore backward-looking; now, with regard to the bad ecological side-effects of chemical methods, they were seen as innovative because of their environmentally protective aspects. Neglected for so many years, biotechnology suddenly seemed to be superior to chemical synthesis. For a long time disregarded as a mere niche technology, it was now perceived even by chemists as a key technology of the future (Bud 1988: 441–4 and 1991: 441–6).

This triggered a new selection process between the alternative technologies, which, this time, turned out to be beneficial for the erstwhile 'loser technology'. But the hasty entry into this field was bound to be a failure from the very beginning. As a consequence of the previous sustained path-dependency, the industrial know-how in biotechnology was inadequate, the gaps in biotechnological research and development were too wide, and the number of biotechnologically trained scientists by far too small (Buchholz 1979: 93). Only now did the drawbacks which the development of biotechnology in West Germany brought with it become evident. Here the chemical-pharmaceutical industry had been far slower in developing and implementing biotechnology than in other highly industrialized countries. Unlike the situation in the United States and Japan, in West Germany biotechnology had found a firm footing neither as an object of research in the universities nor as a production technology in industry. The resulting lag was striking. While there were about 300 scientists working in West Germany in biotechnological R&D in 1970, one could find 3,000 at the same time in Japan (Buchholz 1979:

70). Even in 1983, out of the 448 newly discovered drugs worldwide deriving from micro-organisms, only four had been developed in West Germany (Conzelmann 1989: 40). While on the one hand the German chemical-pharmaceutical industry was still the world leader in chemical synthesis, it was on the other hand a laggard in biotechnology.[26]

Summary and Outlook

To sum up, the backward state of biotechnology in Germany, which became striking in the 1970s, was the result of a long path-dependency on chemical synthesis. The shaping of the path happened gradually and was assured by the interaction of several factors. As the survey shows, already during the Weimar Republic the chemical dyestuff industry had decided in favour of chemosynthesis and thus against biotechnology. However, its further expansion was considerably intensified and accelerated through the politically motivated intervention of the state during the Third Reich. The lasting bias in favour of chemical synthesis ultimately caused the predominance of the German chemical-pharmaceutical industry in this sphere. The costs of this development, however, were to the detriment of biotechnology. In Germany it stood more deeply and for a longer time in the shadow of chemical synthesis than in other highly industrialized countries.

Meanwhile, due to massive public and industrial efforts, a close-meshed biotechnological research infrastructure has been established in Germany. Universities offer programmes in biotechnology, and many enterprises of the chemical-pharmaceutical industry are engaged in biotechnological production. In particular the results of molecular biology and the methods of genetic engineering have enlarged the possibilities of biotechnology to a broad extent. Nowadays the knowledge of the genetic architecture of micro-organisms holds out the prospect of devising concrete strategies of innovation. Biotechnology ranks today among the most research-intensive and science-based technologies (Giesecke 1998: 35). According to several predictions the situation in a so-called 'synthetic biology' will lead to a development comparable to that in synthetic chemistry a hundred years ago. Therefore it might initiate a similar impulse for industry (Winnacker 1990). It seems that the former underestimation of the subject has meanwhile turned into its opposite.

Indeed, there are already some signs indicating the shaping of a new technological path, now in favour of the former 'loser technology'. The dynamics of this process can be illustrated by the example of Hoechst. This company, which is one of the oldest enterprises of the synthetic dyestuff industry and therefore has a tradition of more than hundred years of chemical synthesis, started some years ago to profile itself as a 'life sciences' company. By the end of 1998, its chief executive officer Jürgen Dormann announced that it was selling the remnants of the old chemistry section of the firm in order to concentrate wholly on the new core business of life sciences, which means: on biotechnology. According to Dormann, chemistry is too dependent on business cycles and therefore will have no future (Schmidt 1998: 106). Whether the high hopes put in biotechnology at the end of the twentieth century will actually be fulfilled, however, remains to be seen.

Notes

1. The following paper is based on the results of my doctoral thesis (Marschall 2000).
2. The shift of the German chemical industry from coal-based to petroleum-based chemistry after the Second World War is the subject of Stokes (1994).
3. Since the nineteenth century, the idea was prevalent in Germany that technological progress would be effected by the scientification of technology. For remarks on the German ideal of scientific technology see Radkau (1989: 40–5).
4. 'High technology' here serves as a label of solutions on a high technological level. As an example consider the coal-based synthesis of fuel which was very expensive, both technically and in energy consumption. The creation of 'high technology' is considered by several authors to be an important feature of the German style of technology. See e.g. Radkau (1989: 16) and Manske (1995: 106).
5. For the use of trajectory models in the history of technology see Belt and Rip (1987), Weingart (1989: 8–14) and Rammert (1992).
6. In this chapter I will concentrate solely on the industrial side of the development. For the background to the delayed establishment of microbiology and biotechnology at the German universities see Marschall (2000: 25–48).

7. The actual products of biotechnology mainly consist of chemically complex drugs, which are already effective in small doses and are produced on a small scale. Examples are given in Weber (1999).

8. After its big crises in the first half of the nineteenth century, agricultural production in Germany increased continuously until the outbreak of the First World War. According to Kiesewetter, after 1848/9 the growth of German agriculture was extreme, especially with respect to potatoes. Their share in agricultural production rose from 1.5 per cent around 1800 to 13.6 per cent in 1913 (Kiesewetter 1989: 154–5). More statistics on agricultural development from 1800 to 1914 are provided by Kiesewetter (1989: 153–60) and Aubin/Zorn (1976: 495ff.).

9. On the history of the German dyestuff industry see e.g. Haber (1958 and 1971), Beer (1958) and Landau/Rosenberg (1992). The development of the pharmaceutical industry is described by Vershofen (1949–58), Liebenau (1988) and Possehl (1989).

10. Thus knowledge of the chemical structure of the final product was a precondition of the theoretical and practical development of new methods of synthesis. See Tanner (1997) and Straumann (1995: 19–52).

11. Unsurprisingly, Carl Bosch decided with regard to the production line of varnish, which was comparably underdeveloped, that such purely empirical domains should be left to firms which are already working in a predominantly empirical way. See Radkau (1989: 267) and Marschall (2000: 160–85).

12. Long into the nineteenth century, the education of pharmacists (or the teaching of the so-called *Apothekerkunst* (art of pharmacy) was akin to that of craftsmen. On this, see Schneider (1972: 180).

13. In a Festschrift of the Institut für Gärungsgewerbe und Stärkefabrikation zu Berlin, published in 1925, the post-war era was described as a period of 'languishing at an accelerating rate'. Owing to the shortage of agricultural raw materials the government issued some regulations for the use of raw materials, which e.g. restricted the use of potatoes for production of alcohol until the mid-1920s. The use of grain for that purpose was still forbidden in 1925. See Institut für Gärungsgewerbe und Stärkefabrikation zu Berlin (1925: 12, 225–7)

14. With respect to the field of lactic acid, the director of the biotechnological production of lactic acid at Boehringer Ingelheim used the metaphor of seven rich years of production before the First World War and seven poor years after the war (Dehio 1966: 1). See also the case study of the biotechnological production of lactic acid at Boehringer Ingelheim in Marschall (2000: 203–66).

15. All the factors mentioned in the main text constitute what Hughes calls 'Technological Momentum'. They were the driving forces behind the further development of catalytic high pressure synthesis in the years after 1918. See Hughes (1969) and Landau/Rosenberg (1992).

16. See Hitler's memorandum on the economic situation and future economic policies from 1936, partially quoted in Petzina (1968: 48–53).
17. This is evidenced by the statements of several contemporary chemists such as Albrecht Schmidt, Carl Oppenheimer or Konrad Bernhauer who all considered biotechnological production to be more effective (Schmidt 1934: 386, 727; Bernhauer 1936: iii–iv).
18. The most important process and product innovation in the realm of biotechnology from the time of the Second World War was the industrial production of penicillin. But the development characteristically took place not in Germany but in the United States. See Pieroth (1992).
19. He wrote:

> The growth of the economy that took place in Germany from 1933 very quickly led to raw material shortages, which made it necessary to direct into the constantly growing production processes all the reserves of raw materials that could be found in the country. The recourse to old material became ever more the necessity of the hour. Decisive for the activities in the field of recycling of old material was the condition that consideration of the national economy took precedence in the economic policy of Germany. A higher level of activity was guaranteed in those areas where there was no promise of profit. (Ungewitter 1938b: 164)

20. According to the CIOS Evaluation Report No 22, File No. XXIX-4 (1946: 10–11) there were food yeast factories in Wolfen, Dessau, Tornesch, Holzminden, Mannheim (Zellstoffe Fabrik, Waldhof), Kelheim-Dornau bei Regensburg, Odermünde bei Stettin, Stockstadt bei Aschaffenburg, Wildeshausen bei Kassel, Mannheim-Rheinau and Regensburg-Schwabelweis.
21. A comprehensive description of the process can be found in Marschall (2000: 308–13).
22. On this see the case studies on various firms in Marschall (2000: 203–349).
23. Many examples of this are provided in the case studies in Marschall (2000: 203–349).
24. For instance, Boehringer Ingelheim stopped the biotechnological production of lactic acid in 1972 (Benninga 1990: 417).
25. In 1977, for instance, there appeared an article in *Die chemische Industrie* which stressed the importance of biotechnology for the protection of the natural environment, the use of sustainable raw materials and the recycling of organic waste. It was argued that with respect to the new economic and ecological situation one could no longer neglect biotechnology, for it was ecologically superior to chemical methods (Rehm 1977: 641–4). See also the preface to the DECHEMA monograph of 1973,

where the authors discussed whether complicated steps of chemical processing could be carried out more elegantly, carefully and efficiently with the aid of micro-organisms or enzymes, which new chemicals compounds could be produced by micro-organisms, and which reactions or products could be performed or achieved on a technical scale by means of vegetable or animal cells (DECHEMA 1973: preface).

26. There are plenty of indications of the backward position of Germany with respect to biotechnology, see e.g. DECHEMA (1974: x), Buchholz (1979: 64), Gassen and König (1994: 3); Gebhardt and Giesecke (1997: 80). For empirical evidence see Schell and Mohr (1995: 10) and Giesecke (1998: 58–68).

References

Aubin, Hermann and Zorn, Wolfgang (eds) (1976), *Handbuch der deutschen Wirtschafts- und Sozialgeschichte*, vol. 2, Stuttgart: Union-Verlag.

Beer, John J. (1958), 'Coal Tar Dye Manufacture and the Origins of the Modern Industrial Research Laboratory', *Isis*, 49: 123–31.

Belt, Henk van den and Rip, Arie (1987), 'The Nelson-Winter-Dosi Model and Synthetic Dye Chemistry', in Wiebe E. Bijker, Thomas P. Hughes and Trevor J. Pinch (eds), *The Social Construction of Technological Systems*, Cambridge: MIT Press, 134–58.

Benninga, Harm (1990), *A History of Lactic Acid Making*, Dordrecht: Kluwer Academic Publishers.

Bernhauer, Konrad (1936), *Gärungschemisches Praktikum*, Berlin: Springer.

Braun, Hans-Joachim (1992), 'Konstruktion, Destruktion und der Ausbau technischer Systeme zwischen 1914 und 1945', in Hans-Joachim Braun and Walter Kaiser (eds), *Energiewirtschaft, Automatisierung, Information*, Propyläen Technikgeschichte, vol. V, Berlin: Propyläen, 11–282.

Buchholz, Klaus (1979), 'Die gezielte Förderung und Entwicklung der Biotechnologie', in Wolfgang van den Daele, Wolfgang Krohn and Peter Weingart (eds), *Geplante Forschung*, Frankfurt a. M.: Suhrkamp, 64–116.

Bud, Robert (1988), 'Great Expectations – the Tale of Biotechnology', *Chemistry in Britain*, 24: 441–6.

—— (1991), 'Biotechnology in the Twentieth Century', *Social Studies of Science*, 21: 415–57.

—— (1993), *The Uses of Life*, Cambridge: Cambridge University Press.

C.I.O.S. Evaluation Report (1946), *Fodder Yeast Plants I.G. Farben-industrie, Wolfen*, Item No. 22, File No. XXIX-4, London: H.M. Stationery Office.

Conzelmann, Claus (1989), *Die neue Genesis. Biotechnologie verändert die Welt*, Frankfurt a. M.: Ullstein.

DECHEMA (ed.) (1973), *Technische Biochemie*, Weinheim: Verlag Chemie.

—— (ed.) (1974), *Biotechnologie. Eine Studie über Forschung und Ent-wicklung - Möglichkeiten, Aufgaben und Schwerpunkte der Förderung*, Frankfurt: DECHEMA.

Dehio, Walter (1966), *Die Industrie der Weinsäure, Zitronensäure und Milchsäure*, Ingelheim: internal publication of the archive of Boehringer Ingelheim.

Delbrück, Max (1902), 'Die Mikroorganismen in ihrer Anwendung auf chemische Umsetzungen', *Zeitschrift für angewandte Chemie*, 15: 693–9.

Dellweg, Hanswerner (1974), 'Die Geschichte der Fermentation – Ein Beitrag zur Hundertjahrfeier des Instituts für Gärungsgewerbe und Biotechnologie zu Berlin', in Institut für Gärungsgewerbe und Biotechnologie zu Berlin (ed.), *100 Jahre Institut für Gärungsgewerbe und Biotechnologie zu Berlin 1874–1974*, Berlin: Westkreuz-Druckerei, 17–41.

—— (1982), 'Von der Spiritusfabrikation zur Biotechnologie', in Versuchs- und Lehranstalt für Spiritusfabrikation und Fermen-tationstechnologie in Berlin (ed.), *125 Jahre VLSF*, Berlin: Oranien-druck GmbH, 13–28.

Dosi, Giovanni (1982), 'Technological Paradigms and Technological Trajectories', *Research Policy*, 11: 147–62.

Gassen, Hans G. and König, B. (1994), 'Perspektiven einer biologisch orientierten Industrie in Deutschland. Die Entwicklung in den USA als Leitbild', *Kontakte*, 1: 3–12.

Gebhardt, Christiane and Giesecke, Susanne (1997), 'Die Spezifität der Entwicklungspfade in der Biotechnologie und der Künstlichen Intelligenz', *Comparativ*, 3: 76–97.

Giesecke, Susanne (1998), *Die Triplehelix von Technologie, Markt und Staat. Innovationssysteme in der pharmazeutischen Biotechnologie*, Berlin: PhD-Thesis.

Haber, Ludwig F. (1958), *The Chemical Industry during the Nineteenth Century. A Study of the Economic Aspect of Applied Chemistry in Europe and North America*, Oxford: Clarendon Press.

—— (1971), *The Chemical Industry 1900–1930. International Growth and Technological Change*, Oxford: Clarendon Press.

Hansen, Emil (1897), 'Vorwort', in Franz Lafar, *Technische Mykologie*, Jena: Verlag von Gustav Fischer, i–iv.

Hughes, Thomas P. (1969), 'Technological Momentum in History: Hydrogenation in Germany 1898-1933', *Past and Present*, 44: 106–32.

Institut für Gärungsgewerbe und Stärkefabrikation zu Berlin (ed.) (1925), *Festschrift zur Feier des 50jährigen Bestehens am 29. September 1924*, Berlin: Verlagsbuchhandlung Paul Paray.

Jasanoff, Sheila (1985), 'Technological Innovation in a Corporatist State: The Case of Biotechnology in the Federal Republic of Germany', *Research Policy*, 14: 23–38.

Kiesewetter, Hubert (1989), *Industrielle Revolution in Deutschland, 1815–1914*, Frankfurt a. M.: Suhrkamp.

Landau, Ralph and Rosenberg, Nathan (1992), 'Successful Commercialization in the Chemical Process Industries', in Nathan Rosenberg, Ralph Landau and David C. Mowery (eds), *Technology and the Wealth of Nations*, Stanford: Stanford University Press, 73–119.

Liebenau, Jonathan (1988), 'Ethical Business: The Formation of the Pharmaceutical Industry in Britain, Germany, and the United States before 1914', *Business History*, 30: 116–30.

Manske, Fred (1995), 'Stärken und Schwächen des "deutschen Technikstils" – Überlegungen zu einem international vergleichenden Forschungsprogramm', in Helmuth Rose (ed.), *Nutzerorientierung im Innovationsmanagement*, Frankfurt a. M./New York: Campus, 103–22.

Marsch, Ulrich (2000), *Zwischen Wissenschaft und Wirtschaft. Industrieforschung in Deutschland und Großbritannien 1880 bis 1936*, Paderborn: Schöningh.

Marschall, Luitgard (2000), *Im Schatten der Synthesechemie: Industrielle Biotechnologie in Deutschland 1900–1970*, Frankfurt a. M./New York: Campus.

Nelson, Richard R. and Winter, Sidney (1977), 'In Search of a Useful Theory of Innovation', *Research Policy*, 6: 36–76.

Petzina, Dieter (1968), *Autarkiepolitik im Dritten Reich. Der nationalsozialistische Vierjahresplan*, Stuttgart: dva.

Pieroth, Ingrid (1992), *Penicillinherstellung*, Stuttgart: Wissenschaftliche Verlagsgesellschaft.

Plumpe, Gottfried (1990), *Die I.G. Farbenindustrie AG. Wirtschaft, Technik und Politik 1904–1945*, Berlin: Duncker & Humblot.

Possehl, Ingunn (1989), *Modern aus Tradition. Geschichte der chemisch-pharmazeutischen Fabrik E. Merck Darmstadt*, Darmstadt: E. Merck.

Radkau, Joachim (1989), *Technik in Deutschland*, Frankfurt: Suhrkamp.

Rammert, Werner (1992), 'Entstehung und Entwicklung der Technik: Der Stand der Forschung zur Technikgenese in Deutschland', *Journal für Sozialforschung*, 2: 177–209.

Rehm, Hans-Jürgen (1977), 'Biotechnologische Verfahren in der chemischen Technik', *Die Chemische Industrie*, 29: 641–4.

Schäfer, G. (1907), 'Die Fortschritte auf dem Gebiete der Milchsäurefabrikation', *Chemische Zeitschrift*, 6: 177–80 and 189–91.

Schell, Thomas von and Mohr, Hans (eds) (1995), *Biotechnologie – Gentechnik. Eine Chance für neue Industrien*, Berlin and Heidelberg: Springer.

Schmidt, Albrecht (1934), *Die industrielle Chemie in ihrer Bedeutung im Weltbild und Erinnerungen an ihren Aufbau*, Berlin: de Gruyter.

Schmidt, Regine (1998), 'Hoechst + Rhône-Poulenc = Aventis', *Deutsche Apotheker-Zeitung*, 50: 106.

Schneider, Wolfgang (1972), *Geschichte der pharmazeutischen Chemie*, Weinheim: Verlag Chemie.

Stokes, Raymond G. (1994), *Opting for Oil: The Political Economy of Technological Change in the West German Chemical Industry, 1945–1961*, New York: Cambridge University Press.

Straumann, Tobias (1995), *Die Schöpfung im Reagenzglas. Eine Geschichte der Basler Chemie (1850–1920)*, Basel and Frankfurt a. M.: Helbing & Lichtenhahn.

Tanner, Jakob (1997), 'Medikamente aus dem Labor', in Thomas Busset, Andrea Rosenbusch and Christian Simon (eds), *Chemie in der Schweiz. Geschichte der Forschung und der Industrie*, Basel: Christoph-Merian-Verlag, 117–46.

Ungewitter, Claus (1938a), *Chemie in Deutschland*, Berlin: Junker & Dünnhaupt Verlag.

—— (1938b), *Verwertung des Wertlosen*, Berlin: Wilhelm Limpert-Verlag.

Vershofen, Wilhelm (1949-58), *Die Anfänge der chemisch-pharmazeutischen Industrie. Eine wirtschaftshistorische Studie*, vols. I–III, Berlin, Aulendorf: Deutscher Betriebswirte Verlag.

Weber, Christiane (1999), 'Gentechnisch hergestellte Medikamente werden immer wichtiger', *Deutsche Apotheker-Zeitung*, 139: 54–6.

Weingart, Peter (ed.) (1989), *Technik als sozialer Prozeß*, Frankfurt a. M.: Suhrkamp.

Welsch, Fritz (1981), *Geschichte der chemischen Industrie*, Berlin: Deutscher Verlag der Wissenschaft.

Winnacker, Ernst-Ludwig (1990), 'Synthetische Biologie', in Jost Herbig and Rainer Hohlfeld (eds), *Die zweite Schöpfung*, München: Hanser, 369–85.

CAY-RÜDIGER PRÜLL

Pathology and Politics in the Metropolis, 1900–1945: London, Berlin and the Third Reich

Introduction

Since the 1970s, it has become well known, as a result of research in science theory and the history of medicine, that social factors and therefore political attitudes affect the work of the scientist and the physician (Moraw 1988; Pickstone 1992; Stichweh 1994). Without doubt, medicine is a part of the British and German national cultures (Payer 1989). The following chapter will demonstrate how the way in which the field of pathology was organized practically and theoretically was directly related to the national political context and those scientific and health care organizations which already existed. The definition of the term 'political' here is not restricted to an action made in order to accomplish certain goals within government, for example in the role of party member. In modern political science, the term is also understood as an action or an influence determining the arrangements of public life, and this is how I shall use the term in this essay. Thus it is possible to view the mental attitudes of physicians and, in our case, pathologists as a part of the political culture of society in general (Sellin 1978).

This essay aims to analyse the history of the discipline of pathology in the cities of Berlin and London in the first half of the twentieth century (1900–45) (for an introduction to the history of 'pathology', see Maulitz 1993). I will concentrate on the pathologists as a social group with their own attitudes and scientific ideas. There will be a special focus on the Third Reich. The Nazi period had a strong impact on the history of medicine in Germany (Kater 1989; Bleker/Jachertz 1993).[1] Because it is otherwise impossible to understand the different aspects of this topic, it is important to have a look at the 'prehistory' of the Nazi period. Furthermore, to understand the German pecu- liarities it is necessary to look beyond national borders. In this essay, I shall use the comparative approach. Above all, this method allows us to demonstrate that the work of pathologists in Berlin and London was part of their respective cultural settings. Pathology in Berlin was burdened by certain developments within its specific scientific culture and organization. Because of this burden, the alignment of the practitioners of pathology with the new regime after 1933 was not a matter of chance (Haupt/Kocka 1996; Sauerteig 1998). Contrasting the situation in Berlin with that in London enables us better to analyse the German predisposition towards the rise of National Socialism. The cities of Berlin and London are well suited for such comparative analysis because each was the main metropolis of its country, and they were also the places where the trendsetters of scientific and political development were located in the first half of the twentieth century. The place of pathology in Berlin and London also mirrored the general development of medical science, and the general political development in Germany and England, respectively (Alter 1993, esp. 11).

First, I will give a short description of the work of the pathologist and an outline of the history of pathology in Berlin and London. Secondly, the organization of pathology in both cities will be analysed against the background of the public health systems of Berlin and London and their respective health policies. In the third part of the chapter, I will deal with the role of pathology, politics and the Third Reich. Finally, there will be a summary and a conclusion.

The History of Pathology in London and Berlin

Pathology as a discipline developed from 1800 and played a major role in the emergence of scientific medicine. It was based on the method of autopsy, i.e. the opening of the human corpse in order to detect the cause of death, and the comparison of the results of the post-mortem examination with the records of the living patient (Paris School of Medicine). Both were done by the physician who had treated the patient. The morgues of many hospitals began to be used as autopsy rooms where physicians could readily compare clinical and post-mortem findings (Ackerknecht 1967; Lawrence 1993). Around 1850, pathological anatomy became an integral part of the conceptualization of scientific medicine in Germany. The Berlin pathologist Rudolf Virchow (1821–1902) was foremost in championing morbid anatomy as one of the most important cornerstones of scientific medicine. Virchow received the first chair at the first Pathological Institute in Berlin (at the Charité Hospital) in 1856, and he gave autopsy a threefold function: first, to detect the cause of death; secondly, as a topic of medical education; third, as the basis of scientific research. Morbid anatomy was based on his principle of 'cellular pathology' (Maulitz 1993). The microscope made it possible to take organ and tissue pathology one step further. The cell was now seen as the smallest unit, where a disease was located (Ackerknecht 1957; Bracegirdle 1993). One important goal of pathology based on this theory was to detect and describe pathologically affected parts of the body. Virchow's style of experimentation included the comparison of static impressions of cells and histological samples under monocular enlargement, but not the examination of the disease process itself (Schmiedebach 1993, esp. 121–2). The autopsy became increasingly important and morbid anatomy supported clinical medicine, i.e. in the repair of a defective organ, for example when removing it with the help of surgery (Tröhler 1993). In Germany, certain physicians became specialists, so-called 'pathologists', and the discipline of 'pathology' was institutionalized at German universities as 'pathological anatomy', based on the method of autopsy. Because of its specialized function, pathology in Germany underwent a process of separation from clinical practice. Until about 1900, every German university established its own chair of pathology and an associated institute (Hort 1987; Pantel/Bauer 1990).

In contrast, specialization in England developed much more slowly. In the case of pathology, the English medical profession continued to advocate strongly the methods and organizational structure of the Paris School. In the nineteenth century, the physician himself, rarely a well-trained pathologist, performed the autopsies of his patients. This was especially true in London, where the autopsy method was mainly recognized at the so-called 'voluntary hospitals' (Maulitz 1987: 109–223). To work as a part-time 'pathologist' there was only an attractive proposition because it enabled the physician to work his way to the top of the ranks, and at best to achieve a post as a consultant physician or surgeon. Around 1900, pathology in London was not professionalized. It was in the hands of physicians, and it was not an independent special discipline (Prüll 1999a: 42–8; Foster 1983).

After the turn of the century, pathology was more commonly used in everyday medical diagnosis and treatment. In Berlin, the discipline – originating at the Virchow Institute at the Charité – spread to the city hospitals, where fifteen smaller pathological institutes and departments were founded. These were organizational and program-matical copies of the Charité Institute, their mother institution. They were headed by a university-trained pathologist, a so-called *Prosektor*, who performed routine autopsies for the hospitals (Gruber 1949, esp. 172–89). In London, pathology as such was now institutionalized at the twelve voluntary hospitals and their medical schools. As in Berlin, pathological institutes or departments were founded (see table 1) (Prüll 1999a: 82–130):[2]

Table 1   Foundation of pathological institutes at the Berlin city hospitals and at the London voluntary hospitals, 1890–1935

	Berlin	London
1890–1900	3	3
1900–1920	8	9
1920–1945	4	1*

* 1934 Pathological Department of the Postgraduate Medical School, Hammersmith Hospital, London

In contrast to Berlin, the departments in London had a different orientation. They followed the new concept of 'clincial pathology', which was not restricted to the performance of autopsies, and moreover relied on the examination of body fluids of the living patients. Its orientation was pragmatic and concentrated on the direct fight against disease. Although the work of clinical pathology was based on the laboratory, it also took place in the wards. Pathological anatomy played only a subordinate role. Other fields became much more important for the work of the clinical pathologist: bacteriology, pathological chemistry, and pathological physiology or experimental pathology. The latter in particular was based extensively on animal experimentation. English clinical pathologists laid strong emphasis on experimental pathology. In the last three decades of the nineteenth century, when visiting Germany (the model for scientific medicine) they did not choose Virchow as their main teacher. Instead, the Leipzig pathologist Julius Cohnheim (1839–1884) became an admired figure in the social community of clinical pathologists in London. Cohnheim favoured functional research. He analysed processes and not static cell-images, and he employed experimental methods. Until 1918, most of the voluntary hospitals in London were equipped with a pathological laboratory and with a clinical pathology unit. This was the basis for increased professionalization of the discipline in London during the inter-war period (Foster 1961; Prüll 1999a: 82–128, 161–70, 171–225).

The emergence of clinical pathology in London from about 1900 corresponded with the rise of a new period of scientific medicine. Physicians now partially abandoned the analysis of morphological structures. Pathological anatomy had passed its zenith, because the human body had been so thoroughly investigated. It was the measurement of the body's functions in the laboratory, and the measurement of the dynamics of certain processes, which was of interest now (Klasen 1984: 17–27). The institutionalization of the new concept of clinical pathology corresponded well with this idea, and London pathologists now profited from lagging behind their colleagues in Berlin. In contrast, Berlin pathology remained indebted to Virchow's heritage and restricted its work mainly to morphological pathologic anatomy. Berlin pathology, even after 1900, was focused on the morgue and not on the laboratory (Prüll 1999a: 131–61). There were only limited efforts to install laboratory sciences at the Charité Institute during the time of Virchow's successors Johannes Orth (1847–1923) and Otto Lubarsch (1860–1933). It is noteworthy that these new sciences were represented mainly by Jewish medical scientists. Three out of six departments, those which represented the laboratory-sciences (bacteriology, experimental biology and chemical pathology), were closed down by Lubarsch's successor Robert Rössle (1876–1956) between 1929 and 1933 (Prüll 1999a: 49–81, 184–223; David/Krietsch 1989) Since the turn of the century, Orth and Lubarsch had already tried to compensate for these inadequacies by giving the backward pathological anatomy a new theoretical basis. They called for a stronger recognition of constitutional pathology (Orth 1904: 1–32; Lubarsch 1927). This theory, which took different forms, appeared throughout medicine, and especially through the entire discipline of pathology, in the inter-war period. Criticism of modern civilization inspired the rejection of materialism and organicism. Constitutionalists favoured the recognition of external (i.e. environmental and social) factors, and furthermore the recognition of internal factors (i.e. disposition and heredity). The whole human being was to be considered under this holistic concept (Klasen 1984; Engelhardt 1985; Weisz/Lawrence 1998).

To summarize, we can describe two different concepts of pathology in Berlin and London, which strongly influenced the discipline in both cities during the first half of the twentieth century. Pathologists in Berlin were advocates of the older morphological concept of Virchow, which was the basis of the foundation of scientific medicine in the nineteenth century. Pathologists in London favoured the new concept of clinical pathology, which was oriented towards new demands which were being made on scientific medicine (Prüll 1999a).

The Metropolis, Public Health and Pathology – Structure and Organization in London and Berlin

Against the background described above, pathologists did their work in Berlin and London as part of their respective health care systems. The pathologists of both cities were members of the middle class and had a good university education – the London pathologists, even in the twentieth century, predominantly at Oxford and Cambridge.

The social and economic development of both cities was the background to the fate of the discipline after 1900. London and Berlin were growing cities since the beginning of the twentieth century. In 1889, the L.C.C. (London County Council) was founded as a central government for the city. In 1900, the capital of Britain got twenty-eight new borough councils, and from then on was called 'Greater London' (Young/Garside 1982: 52–101; Beckson 1992). The capital of the United Kingdom was a pluralistic society with many different groups of immigrants. London was dominated very much by the rise of the Labour party, which supported urban centralization – in contrast to the Conservative party with their 'suburban mentality'. But the state seemed to be neutral, and therefore the Conservatives could tolerate the Labour party's dominance in London (Middlemas 1993; Daunton 1993: 99–102; Thompson 1967: 90–111). In the inter-war period, London became the centre of impressive economic growth. The first half of the century was a period of social and economic stabilization (Green 1967, esp. 32–3).[3]

Berlin was also the capital of a nation, the administrative centre of its country, and a European metropolis. But at the turn of the century, Berlin – in contrast to London – was still a young city which had only begun to gain importance with the foundation of the German Empire in 1871. Meanwhile, Berlin had become one of Europe's economic and intellectual centres through progress in manufacturing and science (Ribbe 1990). The growth of the city was symbolized by the unification of the inner districts with the smaller villages in the suburban areas and the creation of 'Greater Berlin' (*Großberlin*) in 1920. In the Weimar period, there was an enormous increase in cultural life, at least between 1925 and 1929, symbolized, amongst other things, by the success of theatres and cinemas. Berlin was the metropolis of the 'Golden Twenties', but, in contrast to London, it was also characterized by a strained social atmosphere. The rising mass culture was not accepted by many, and there was hostility as well as admiration. Between 1918 and 1924, Berlin experienced the greatest political upheaval in its history. It changed from the centre of power of the autocratical Wilhelmine Empire to a democratic city, which was searching in vain for direction under a new form of government. At the end of the First World War, wracked by political controversies and conflicts between militant right-wing and left-wing groups, Berlin was a city rent by struggle. Political conflicts exposed dangerous rifts in society. Even during the period of stabilization between 1924 and 1928, it was not possible to solve these social conflicts – conflicts which had their origins in the different occupations and living quarters of the inhabitants of Berlin. Not everybody benefited from economic prosperity, which in any case came to an end in the worldwide economic crisis of 1929 (Lehnert 1993). Until that year, the parliament of the city was dominated by the democratic parties (SPD, DDP, Zentrum and DVP), but in 1932/3, these parties received the votes of less than 30 per cent of the people of Berlin. In 1931, the government of the city wanted to control the political crisis and strengthened the authority of the Lord Mayor, and thereby partially abolished the city's democratic autonomy (Ribbe 1990: 43–4; Ribbe 1987: 845ff.; Büsch/Haus 1987: 358–60). Civil servants were hostile towards the democratic state, as were the army, the students, and especially the university teachers. The University of Berlin remained an advocate of the lost Empire (Reich), its members showing hostility toward the metropolis as an incarnation of the new mass society and the bad influences of civilization (vom Brocke 1990, esp. 178–9; Anselm 1987, esp. 255). Between 1933 and 1945, Berlin was the capital of Nazi Germany, and this was the final step of a process characterized by destabilization (Steinbach 1990).

In the inter-war period, both London and Berlin saw an enlargement of their respective public health care organizations. This was aimed at an improvement of maternal and child health care, and also at the prevention of infectious diseases, i.e. tuberculosis or syphilis. In this programme, both cities led their respective countries. The public health care organizations corresponded to the general political structures in both cases: in London, we can only trace slow progress in health care reform measures, which were decentralized because of the responsibility of city districts and boroughs. In contrast, Berlin saw a rapid centralization of the field of public health (Weindling 1993).

In London a long tradition of ecclesiastical and philanthropic initiatives as the main backing of the health care system prevailed. Innovations could be introduced mainly within the framework of old, religiously motivated institutions, and medical specialization – as in Berlin – took place very late. In London, health care relied chiefly on the private voluntary hospitals, which worked independently and which were not subordinated to any government or urban regulations. In contrast, Berlin saw the development of a modern hospital system on the basis of government influence during the first half of the twentieth century. Financing of city hospitals was a cornerstone of autonomous urban health care policy.

The end of the 1930s brought a change of health care policy in Berlin and London: the British metropolis saw the rise of power of the L.C.C., which controlled the hospitals of the so-called 'Asylums Board' through the Local Government Act of 1929. This meant the exclusion of a huge number of sick persons from the poor law and their integration into the urban health care system. Since the beginning of the 1930s, the Labour party had dominated the work of the L.C.C. These changes resulted in a more effective urban self-government. But it was still decentralized, because private voluntary hospitals remained independent institutions. In contrast, the National Socialist rise to power in Berlin in 1933 suddenly put a halt to the tremendous increase in urban self-government and ambitious reform plans. The health care system was now organized according to the 'leadership principle' (*Führerprinzip*), and this intention was facilitated by the existing centralized governmental structures of the German metropolis. To sum up: the differences between the histories of both cities are mirrored in their different public health care systems (Weindling 1993: 232–3; Berridge 1990).

Additionally, the development of pathology corresponded with the development of health care systems in both cities. The pathological departments of the London voluntary hospitals were not controlled by any one institution, and all twelve represented a decentralized network. Decentralization and the varying levels of influence of the respective hospital governments and administrations of the medical schools determined the size and type of the institutions. Sometimes only one subdiscipline of clinical pathology was represented under the leadership of one pathologist, while at other times several units existed on an equal level, and in certain cases one director managed a big institute with several semi-independent units. In contrast, the sixteen pathological institutes at the Berlin hospitals were oriented towards the Pathological Institute of the Charité Hospital, and their heads were dependent on the director of the latter. It was a centralized system. The leading pathologist at the Charité Institute paid great attention to the appointment of hospital pathologists. These positions were seen as the first step towards a professorial career. The director of the Charité Institute tried to influence these appointments, although the final decisions were made by the municipal authorities. On the other hand, if the hospital pathologists were interested in working in a good atmosphere, they had to be on good terms with the director of the Charité Institute. The heads of the Pathological Institute of the Charité were keen to avoid the independent founding of any new pathology departments or institutes by clinicians (Prüll 1999a: 65–81, 128–30, 227–55).

Even the organization of the individual institutes or departments in both cities showed the difference between a centralized and a decentralized structure, which was strongly influenced by the respective scientific and educational systems. Pathology in Berlin was determined by autocratic structures, which had emerged parallel to the hierarchical political system of Imperial Germany. These structures were characterized by the powerful position of the leading professor (*Ordinarius*) of pathology at the Charité Hospital, who conducted his own negotiations with the Prussian Ministry of Science or other colleagues on his social level. It was only possible to influence the working or structural conditions in the Pathological Institute via the professor, whose authority could not be violated.[4] In contrast, London institutions of pathology were much more open to influences from the hospital, the medical schools or lay people. Clinicians were able to participate in organizing pathology, because any decision-making in London voluntary hospitals took place through committees. Even the head of a pathological unit was sometimes answerable to such a committee. The pathologist had to present a report of his work at regular intervals, and had to submit to evaluation. Thus, London pathology was much more transparent than its Berlin counterpart, and the demands and interests of society carried a much greater weight. London pathology could influence the pathologists through the lay governments of the hospitals and the subordinate committees, and therefore co-determine research, education and the routine work within the discipline.

We can conclude that pathological anatomy in Berlin and clinical pathology in London were organized in different ways, which in each case corresponded with local policy in general, and with health care policy in particular. London pathology was part of a stable but decentralized system. Based in the voluntary hospitals, clinical pathology relied mainly on the demands of the private market. In contrast, Berlin pathology was part of an unstable, centralized system under political pressure, which was driven mainly by governmental directives. This was the background for the political challenges of the inter-war period.

Pathology, Politics and the Third Reich

The Age of the Empires and the Pathologists Political Attitudes

Since the turn of the century, pathology in London and Berlin had been tied to the socio-political goals of their respective countries. London pathologists were involved for example in colonial wars, giving medical aid and support. Furthermore, physicians in England saw military duties as a means of improving their social status vis-à-vis their German rivals (Harrison 1996). In Berlin, pathology was embedded in the policy of Imperial Germany, where army physicians were delegated to the Pathological Institute at the Charité. This custom was intensified after 1902, when Johannes Orth, who had connections to the royal family, became the successor to Rudolf Virchow (Fick 1924: xcvi; Hubenstorf/Walter 1994, esp. 26; Orth 1910, esp. 171).

Organizational and scientific links between pathology and politics in London and Berlin became particularly evident in the First World War. Pathological institutes focused their research interests on war wounds and diseases which mainly occur in times of war. Work had to be done under the difficult conditions of war. The departments were ill-equipped in staff and supplies. In 1918, for example, only one out of seven assistants at the Charité Institute in Berlin was still there. In 1916, due to the shortage of material, it was impossible to prepare microscopic specimens at the pathological museum of the Royal Free Hospital in London, and the material had to be preserved for use at a later time.[5]

The work of Berlin and London pathologists in the First World War corresponded to their respective national conceptions of pathology. Furthermore, both capitals were the headquarters for their respective fields. In England, military pathology was organized by Colonel Sir William Boog Leishman (1865–1926), a well-known specialist in tropical medicine. His statements about the major tasks of pathology in the First World War were pretty much a mirror of military pathology's general direction in England. The goal of pathology would be 'to offer aid for the common aim – the maintenance of the health of the troops and the effective treatment of the sick and wounded' (Leishman 1923, esp. 5).

This was the pragmatic and utilitarian concept which aimed to help every soldier exploit all national resources. The work relied on clinical pathology, which means the bacteriological and pathochemical work in 'stationary' or 'mobile laboratories'. Autopsies were only of minor importance (Leishman 1923: 9–12).[6] British military pathology was federalistic in its structure, as Leishman had to answer to committees and held only restricted authority in his field. In contrast, German 'war pathology' (*Kriegspathologie*) was organized autocratically by the pathologist at Freiburg University, Ludwig Aschoff (1866–1942) (Prüll 1999b). It was heavily theory-based and focused on morphological work.

Although German army pathologists investigated suspicious deaths by conducting autopsies, they accepted that providing new insights into constitutional pathology was the main task of war pathology. The overall constitution of the individual should be determined. The behaviour of the healthy, able-bodied male organism and its adaptability to the stresses of war were at the centre of interest. Illness was seen as a sign of inferior quality and degeneration, and corresponding morphological changes should be detected. To gather information about normal anatomy and physiology, an autopsy of almost every soldier had to be performed as soon after death as possible. Research in the field had to yield results not only for clinical medicine, but also for the whole nation, whose fate in the 'cultural war' and its aftermath was being determined (Prüll 1996: 155–82; Aschoff 1916 and 1921). Berlin pathology, especially represented by

Otto Lubarsch, was integrated into this programme, which was backed to a great extent by contemporary ideas about racial hygiene. Lubarsch had launched verbal attacks on France and Britain, who would send 'yellow, black and brown or, if possible, the white scum of the earth' (Lubarsch 1916: 9) against Germany. Under his auspices, army physicians performed animal experiments to test gas, which could be used for chemical warfare.[7]

The Pathologist as Health Care Specialist in London after the First World War

After 1918, pathology in London saw a period of rapid development. Of course, the discipline was not confronted with any general political changes. London had no Third Reich and the severe class conflicts which swept through Europe did not affect England with the same virulence as elsewhere. The country experienced a relatively stable political situation, and many people – fearful of any change – called for the restoration of the old Victorian habits of voluntary action and self-help. But the inter-war period in England saw changes in the relations between government and society, resulting in increased 'government intervention in industry, extension of social welfare schemes and general administrative rationalisation' (Harris 1990, esp. 74ff., quot. 80). The government somehow came closer to society. In November 1927, it was therefore not surprising that the Laboratory Assistants Association would meet for three days at the Bland-Sutton-Institute of Pathology at the Middlesex Hospital in London to organize the representation of their interests.[8] Pathologists in London had the opportunity to concentrate on organizing a laboratory service in the metropolis, and also in England in general. This meant a contribution to health care measures for society. While pathologists in Berlin, because they had restricted their attention to morphology, had only limited or indirect access to public health care, pathologists in London were able to participate in these movements because of their wide understanding of 'pathology'. In respect of the contemporary discussion about the old Victorian values of self-help and competitive free market forces, on one side, and new governmental care and social policy, on the other, London pathologists tended to prefer the former way of thinking. Well-known persons, for example Philip Noel Panton (1877–1950), clinical pathologist of the Royal London Hospital, fought for the private hospitals, which were based on traditional philanthropic thinking and which would prosper when in competition with their own kind. Panton idealized the big London voluntary hospitals when comparing their development with the British people's voluntary organization of a permanent army or of the battle fleet (Panton 1951: 80). Panton protested vehemently against every kind of control of the voluntary hospitals which could be exercised by the L.C.C. Therefore, he was one of the defenders of the voluntary hospital as an elite institution who tried to avoid any influence of general practitioners on the structure and management of the latter (Honigsbaum 1979).

Accordingly, clinical pathology became established at the London voluntary hospitals after 1918. Although there were some shortcomings in the field, it had proved to be fruitful in the First World War.[9] The discipline now had to negotiate with the L.C.C., which itself founded laboratories to secure public health care. In 1929, when the British government had put an end to the Poor Law, the L.C.C. got the supervision over 120,000 hospital beds in the London districts. Only one year later, in 1930, the L.C.C. developed a plan for the foundation of an independent laboratory service. The conditions for the performance of routine examinations at the smaller hospitals were improved, and five new central laboratories were founded to offer support in some of the more difficult tests. Furthermore, the twelve voluntary hospitals were integrated into the scheme, but their pathologists, on the other hand, feared severe competition.[10] The very expensive units of the voluntary hospitals were in danger of being closed down. During the Second World War, clinical pathologists were successful in defending their dominant position in laboratory service against physicians, who favoured socialization of the health care system and the takeover of laboratories by the L.C.C. (Foster 1983: 56–8).

Clinical pathology was also professionalized in the inter-war period. In 1926, the Clinical Association of Pathologists had been founded. Until 1930, the latter was able to secure the foundation of departments of clinical pathology at the private hospitals outside of London (Dyke 1961, esp. 130–5). Professionalization of clinical pathology during this period was further supported by the specialization of British medicine in general. In the inter-war period, disciplines prospered which had been underdeveloped until that point, and which now tried to achieve independence from the dominant fields of 'medicine' and 'surgery' (Moscucci 1993: 181). These efforts, which culminated in the foundation of the Royal College of Pathologists after the Second World War in 1964 (Cunningham 1992: 335–61),[11] took up the entire energy of clinical pathologists until the end of the 1930s.

During the Second World War, clinical pathologists were integrated into new governmental measures for medical care and treatment. In 1940, the laboratories of the voluntary hospitals became a part of the E.M.S. (Emergency Medical Service), which was subordinated to the Ministry of Health and was to provide health care for British people in wartime. London was divided into sectors, and each of the voluntary hospitals did laboratory service for one of these. The head of each Voluntary Hospital Pathology Unit was called the 'Sector Pathologist'. His work was done as part of the G.L.E.L.S. (Greater London Emergency Laboratory Service), and he was a representative of the E.M.S.[12] The sector pathologist had the task to install further E.M.S. laboratories in his sector and had access to the hospitals' therapeutics, for example units of stored blood.[13] He did routine work, but also educated pathologists and conducted research on a restricted scale. For example, James McIntosh (1882–1948), pathologist at the Middlesex Hospital, trained seventeen pathologists during wartime. Furthermore, he gave 'penicillin courses', teaching the correct way of using this new antibiotic, and he performed research on antiseptic and bacteriostatic substances (McIntosh 1944; Windeyer 1945). The latter topic was of interest to society, and it was not by chance that clinical pathologists played an important role in the development of penicillin, which was produced on a largescale from 1942, and which was a result of pragmatic interest in combating disease.[14]

Therefore, London pathologists were successful in advancing their professionalization. Important representatives of the field could achieve key positions at the E.M.S. In 1934, to counter any foreign bacteriological warfare, the British government organized a committee for the purpose of preparing British laboratory organization in the event of war. One member of this committee was William Whiteman Carlton Topley (1886–1944), who in former times had been pathologist at Charing Cross Hospital in London. Topley went on with his work and founded the E.P.H.L.S. (Emergency Public Health Laboratory Service). In 1941, Topley joined the Agricultural Research Council to perform pathological work to ensure an adequate cereal supply. Topley's successor was Philip Noel Panton. As a member of the E.M.S., he organized the laboratory service in England during the last four years of the war. He saw to the employment of qualified personnel at the clinical pathology units of the hospitals and to the improvement and foundation of laboratories. After the war, he was raised to the peerage because of his initiative.[15] The work of Panton contributed decisively to the organization of laboratory medicine in England after 1945, and to the creation of the Public Health Laboratory Service (the successor of the E.P.H.L.S.) as a part of the new NHS (National Health Service), which came into existence in 1946 (Webster 1988: 316–17; Webster 1996).

The Pathologist in Berlin as a Political Theorist: the Weimar Republic and the Nazi Period

In contrast with the situation in London, Berlin pathologists saw political upheaval, but no political stability, in 1918. They were confronted with a society in crisis. In the course of the military defeat, a conflict emerged between the traditional forces of Imperial Germany and the supporters of the new democratic government. This conflict also affected Berlin pathologists. As a traditional discipline which had played an important role in the birth of scientific medicine in Imperial Germany, pathology remained attached to the old regime. Berlin pathologists did not follow the new democratic approach, but tried to promote German nationalist ideas even within science. Aschoff's war pathology became an important part of pathology work in Berlin during the Weimar Republic. After the war, material that had been gathered in wartime was subjected to evaluation within the framework of constitutional pathology. This material comprised organ preparations of soldiers who had died in or after the war. There were two main institutions which continued work in the field of war pathology, and both were located in Berlin. First, and most prominently, there was the Collection of Material on War Pathology and Constitutional Pathology (Kriegs- und Konstitutionspathologische Sammlung) at the *Kaiser-Wilhelms-Akademie* in Berlin. Since the collection's establishment in 1917, it had been in the hands of Aschoff's pupil, Walter Koch (1880–1962), who had been instrumental in organizing war pathology. In 1921, the Berlin Collection boasted 6,000 preparations, a vast registry covering approximately 70,000 dissection reports, and microscopic material (Aschoff 1921: v–vi; Giese 1963, esp. 423). Secondly, the evaluation of this material was partly published in a continuous series entitled *Publications from the Fields of War Pathology and Constitutional Pathology*, established in 1920. The editors of this series were Aschoff and three other former war pathologists. Walter Koch was the secretary of the editorial board.

As with the collection, Aschoff and Koch were the organizers of the series, and in the first volume they outlined its objectives. The scientific experiences of clinicians and pathological anatomists in service during the First World War were to be published here. Accordingly, articles in the publication series were devoted to assessment of the preparations, to dissection protocols, and to the pursuit, either indirectly or directly, of issues of constitutional pathology (Aschoff et al. 1920). The latter was linked with racial hygiene and contained overt political references.

In 1921, Robert Rössle, together with an assistant, conducted a study of the physical growth of schoolchildren. Although internal influences were taken into consideration, the study dealt primarily with external influences on children's growth. Like war, school was deemed an 'experiment on the masses' in which 'individuals with, genotypically speaking, totally different reactions were pressed into moulds that were similar to each other'. Schoolchildren were therefore viewed as a group of people with one specific constitution (Rössle/ Böning 1927, esp. 5).[16] On the one hand, the authors noted that in 1921, despite the First World War, the height of schoolchildren had increased as compared to the height recorded in 1880; this change was termed *Aufartung* (,improvement in the quality of the race'). On the other hand, the authors also argued that school had, on the whole, a deleterious influence on growth. Discordant growth processes were attributed to public education, and in a dig at the new democratic government, to the lack of 'military education', which could have provided things that the school had neglected to teach (Rössle/Böning 1927: 24, 28, quotation: 52, 53).

Such attitudes were increasingly significant in the context of a falling birth rate and extensive war casualties, which provoked fears of the demise, or at least the degeneration, of the German people. The research projects conducted in constitutional pathology during the Weimar period, which took up the thread of war pathology, were right-wing efforts of German pathologists to influence the German state's health policy. In the 1920s, war pathology was therefore adapted to new conditions, and its orientations widened to meet the extended national duties. The Reich Ministry of Labour passed a resolution to expand the War Pathology collection into a scientific department for social and occupational medicine. This was in keeping with the fact that occupational diseases, until then severely neglected in Germany, were now subject to legislation. In May 1925, the Reich's Minister of Labour promulgated a decree in which, for the first time ever, accident insurance coverage was extended to include eleven occupational diseases; in February 1929, this directive was expanded to include twenty-two occupationally induced diseases. The collection subsequently came under the aegis of the Reich's Public Health Office (Reichsgesundheitsamt).[17]

In 1925, Max Busch (1886–1934), previously a university lecturer at the Pathology Institute of Erlangen, succeeded Koch as head of the department of pathological anatomy at the Reich's Public Health Office. Although now concentrated on occupational patterns, the collection continued to pursue the goals of war pathology. War and occupational medicine were intimately linked with each other. Busch, in accordance with his extremely nationalist views, secretly lectured at the School for Chemical Warfare of the Reichswehr (*Gasabwehrschule*) on battle lesions ensuing from chemicals. Secretly, too, the German Railways (Reichsbahn) had its officials trained in chemical warfare. The Reichswehr carried out the training, and Busch conducted the demonstrations. In public exhibitions, constitutional pathology was linked with war-related illnesses as well as occupational diseases.[18] In 1931, correspondingly, the publication series of constitutional pathology changed its title to *Publications on Occupational and Constitutional Pathology*, because 'the effects of the "occupational battle" call for our increased interest on a day to day basis'.[19] The editors transferred their military vocabulary to their new field of work. The term 'war battle' was equated with 'occupational battle'. Whereas, in the past, the soldiers on the battlefront had been the focus for studies of the state of the German people's health, now workers at the labour front were the centre of interest. In the eyes of contemporaries, both groups could be considered the pillars of society, and both worked in areas in which individuals had to prove themselves under extreme conditions. This is illustrated by a novel (The Labourer, *Der Arbeiter*, 1932: esp. 37, 39) by Ernst Jünger, a former officer and assault force commander in the First World War, and one of the most celebrated authors of the inter-war period in Germany. Involved in an interminable process of war, the labourer, in Jünger's view, was the descendant of the soldier of the First World War. His anti-republicanism and his 'heroic nationalism', firing Jünger's hope for a new powerful era, exemplify the mood in which war pathology was turned into labour pathology, yet remained faithful to its original purpose (Sontheimer 1994: 103–6; Meyer 1993: 163–213). Walter Koch put it this way: 'Occupational pathology is actually the war pathology of peace.'[20]

Consequently, during the 1930s, the collection and publication series returned to their original mission of war pathology. In 1936, just three years after Hitler's rise to power, the editors of the series declared that in future the articles would bear the title *Publications on Constitutional Pathology and Military Pathology* 'as the German *Wehrmacht* has again taken its former place in the life of our nation.'[21]

The Berlin Collection was placed under the direction of the pathologist Paul Schürmann (1895–1941), a pupil of Robert Rössle, in 1935, when it fell under the control of the Military Medical Academy of the army. In the same year, Schürmann founded the Institute for General Pathology and Military Pathology as a part of the Academy (*Institut für allgemeine und Wehrpathologie an der Militär-ärztlichen Akademie*) and thus elevated pathology to a basic subject within military medicine.[22] There was a relatively smooth transition from the theoretical and organizational work of German war pathology of the First World War to the era of National Socialism. Because predisposition and constitution were strongly determined by internal factors, such as heredity, there was no major difficulty in linking constitutional pathology to Nazi ideology (Lampert 1991a and 1991b). In September 1933, Walter Koch offered brief courses on heredity and racial biology in the Kaiser Wilhelm Institute of Anthropology in Berlin (Kaiser-Wilhelm-Institut für Anthropologie, menschliche Erblehre und Eugenik). Since 1925, he had been head of the Academy of Social Hygiene in Berlin-Charlottenburg, which was promoting the courses. Koch organized the syllabus and made contact with participants and lecturers from the entire Reich. He maintained close contact with Leonardo Conti (1900–1945), who was appointed 'Reich Health Leader' (Reichsgesundheitsführer) by Adolf Hitler in 1939 (Giese 1963: 424; Prüll 1997b and 1998).23 Walter Koch's nationalism was no exception. His person exemplifies the military-pathologist tradition of the discipline in Berlin. Among twenty-five pathologists who worked at the Berlin hospitals between 1918 and 1933 in leading positions, there were at least seven (28 per cent) with a nationalist mentality hostile to democracy. The number may have been higher in total.[24]

One of the most nationalist pathologists was Otto Lubarsch, the director of the Charité Institute since 1917. Lubarsch was a founding member of the Pan-German League (*Alldeutscher Verband*) as early as 1890. As an admirer of Prussian-led Germany, he signed two annexationist petitions to the Reichstag during the war. After the war, he continued to pursue his political objectives as a member of the

German National People's Party (*Deutschnationale Volkspartei* (D.N.V.P.) and as chairman of the committee of German National University Lecturers (Lubarsch 1931: 531–2, 538–9; Döring 1975: 61, 266). Lubarsch tried to avoid any democratic interventions in the government of his institute. He combated the increasing participation of the assistants in university life and their meetings on the premises of the Pathological Institute. In 1923, Lubarsch was advised by the Charité administration no longer to use the stamp 'Royal Pathological Institute' (Königlich Pathologisches Institut) (Prüll 1997a, esp. 198–9). Furthermore, Lubarsch gave pathological anatomy a nationalist function. For him, pathological anatomy had a cultural meaning and was not only a science but a mentality, concerned with a 'rigid and straight disciplinary orientation of the mind' (Lubarsch 1917, esp. 1380).[25] The discipline had to be strengthened again as much as Germany herself, and Lubarsch compared his efforts to reintegrate bacteriology into pathology with the reclamation of the eastern parts of Germany, which had been lost in 1918 (Lubarsch 1927: 5). He oriented his scientific work towards his political aims and reinterpreted Virchow's 'cellular pathology' on the basis of his constitutionalist ideas. Virchow had compared the cells with parts of a state, and he had attributed to them the important function of central power, while the government would have only coordinative duties. Lubarsch, in contrast, compared illness of the cells with a 'proletarian revolution' which, in case of an escalation, had to be suppressed by the central power. This would be a life-and-death struggle, leading to health or the destruction of the whole body (Lubarsch 1931: 591–2, esp. 591; Prüll 1997a).[26]

Lubarsch was a Jew and had several Jewish assistants. Therefore there was anti-Semitic feeling against the Institute. Walter Koch told Aschoff in 1929 that the Institute under the direction of Lubarsch would become 'Jewish' (*verjudet*) and that he and a colleague would promote the pathologist Julius Wätjen (1883–1968) as new head of the anatomical department to 'preserve the clean mentality' of Lubarsch's predecessor Johannes Orth.[27] Lubarsch retired in 1928, and the provisional head of the Institute between 1928 and 1929, Julius Wätjen, was not only a nationalist but also had anti-Semitic attitudes. In 1929, when Robert Rössle occupied the chair and initiated some changes in personnel, Wätjen described this process as a 'big clean-up session'.[28] Later, from 1 May 1933, Wätjen became a member of the Nazi party and several Nazi organizations, among others a supporting member of the SS (*Schutzstaffel*). Even in 1933/4,

Wätjen called for the replacement of 'academic liberty to teach' (*akademische Lehrfreiheit*) with 'academic liberty to serve' (*akademische Wehrfreiheit*) (Lampert 1991a: 72–3).

Despite these conditions, the National Socialists' rise to power involved a setback for the Charité Institute. In 1933, many members of the academic staff were Jews. Therefore, the new regime viewed the Institute with suspicion. Even in November 1932, the National Socialist Party Cell Organization in Works (Nationalsozialistische Betriebszellenorganisation, NSBO) of the Charité Hospital had spied on the Institute and had presented secret reports to the Science Ministry. In particular the chemical department under Peter Rona (1871–1945) was in its view 'totally in the hands of the Jews'.[29] The NSBO assumed that Robert Rössle, who even after 1933 did not become a party member, would offer resistance to any dismissal of assistants, and thus wanted to 'collect material about the so-called "scientific" work of the Jews'. Certain groups within the Science Ministry furthermore tried to bring it about 'that the first and biggest institute of pathology in Germany would employ German physicians'.[30] Consequently, in 1933, seven out of about fourteen assistants were dismissed, along with the head of the chemical department, Peter Rona. Rössle's assistant medical director (*Oberarzt*), Frédéric Roulet (1902–85), went back to Switzerland. In June 1933, Rössle sent a list of names to the Charité administration, new assistants who would support the 'national state' (*nationaler Staat*) at all times.[31] Overall, the National Socialists' measures caused great scientific and human losses to the Charité Institute. One example is the emigration of Peter Rona and his pupils – among others Ernest Boris Chain (1906–79), who later worked together with Howard Walter Florey (1898–1968) on penicillin in Oxford – which put an end to all his efforts to promote research in biochemistry in Berlin and Germany. Many scientists had received their basic education in biochemistry at Rona's department before they went to the much sought-after institutes of the Kaiser Wilhelm Academy in Berlin (Engel 1994, esp. 308–9). Furthermore, the new assistants were sometimes restricted in their capacity to work because of their compulsory military training as members of the *Wehrmacht*.[32]

This was only one side of the coin. Rössle sometimes tried to help persecuted colleagues (Lampert 1991a: 83–4), but on the other hand adjusted quite well to the new state. He even successfully took advantage of the political changes when strengthening the anatomical department and the morphological work conducted there. The dismissal of Peter Rona and the closing of his department of chemical pathology as an independent unit in 1933 was only the last step in Rössle's programme to centralize the power of the Institute and focus its research and education mainly on the anatomical department, using bacteriology and chemistry only as supporting minor disciplines.[33] The bacteriological and experimental-biological departments had been abolished, because, in Rössle's view, both had worked selfishly and had not participated in dealing with 'questions of pathology'. The result was an extended anatomical department with a budget of 40,000 RM (*Reichsmark*) and a small department of chemistry with a budget of 8,000 RM.[34] Only in the course of the so-called *Gleichschaltung* (equalization) in 1933 could Rössle gain control of the department of chemistry, which until then had been led as a small unit by Karl Hinsberg (*1894). The biochemical approach was abandoned, and Hinsberg devoted his work to the support of the morphological approach from then on (Engel 1994: 310).[35]

The organization of pathology at the Charité under Rössle between 1932 and 1945 corresponded to Virchow's ideas at the turn of the century, and Rössle promoted pathological anatomy much more aggressively than his predecessors when integrating other fields into pathology, fields which had been lost to other disciplines or which were already independent. In 1919, the biologist Rhoda Erdmann (1870–1935) had introduced the method of cell and tissue cultivation in Germany after her return from the United States. In the same year, she was able to found an Institute of Experimental Cell Research (Institut für experimentelle Zellforschung) in Berlin. As Department of Experimental Cell Research (Abteilung für experimentelle Zellforschung) at the Charité Hospital, this institution was regarded as the first of its kind in Germany. In 1935, after the Nazis, rise to power and after Erdmann had died, Rössle was able to exploit a conflict between the two physicians Hans Auler and Heinz Zeiss (1888–1947), who both wanted to control the vacant institution. With the help of his friend Ferdinand Sauerbruch (1875–1951), the famous Berlin surgeon, Rössle was successful in absorbing the Institute of Experimental Cell Research and creating a new, dependent subdepartment at the Pathological Institute (Prüll 1999a: 304–6; Hubenstorf/Walter 1994: 36–7). Cell research changed from an independent area with general research interests to a supporting field of morphological pathology. Any knowledge about growth and reproduction of cells was now applied primarily to one of the old morphology-based research areas – the pathology of cancer. With the subordination of cell research under the auspices of morbid anatomy, other directions were cut off. Even in 1949, the German pathologist Walther Fischer (1882–1969) had to confess that tissue cultivation in German pathology was mainly restricted to questions of cancer research (Engel 1994: 303–4; Fischer 1949).

The institutional concentration on morbid anatomy was accompanied by similar tendencies in Rössle's personal scientific work, which corresponded to the interests of the new government. These interests were deeply rooted in his engagement in the war pathology of the First World War. Rössle tried to evaluate the autopsy material on the basis of constitutional pathology. In 1919, he had illustrated the polarization between the healthy young soldier at the front, who was strengthened by the catharsis of the war, and the 'decrepit human material of the city'[36] that the pathologist would often encounter (Rössle 1919, esp. 34). Although he had noticed its methodological weaknesses, Rössle was a protagonist of the constitutional approach even after 1933. In 1934, he confessed that the field had no method at all (Rössle 1934, esp. 16), but it was important for him to save pathological anatomy. So he devoted himself to questions of heredity, and in particular performed family and twin research when comparing the location and consistency of the human organs. Rössle himself admitted the impossibility of differentiating between endogenous and exogenous factors of a disease with the help of morbid anatomy. On the other hand, he claimed that the recent pathology of heredity was 'one of the most important sources of understanding diseases' (Rössle 1939: 284; Rössle 1940). In 1934, Rössle pleaded for the prevention of 'therapeutically hopeless offspring'.[37] Rössle welcomed the law of compulsory sterilization, passed in the same year, and he saw the state as a protective authority in respect to carriers of supposedly inferior genotypes (Rössle 1934: 26; Lampert 1991a: 83–5). His assistants knew that Rössle was interested in autopsies of twins and relatives and that he collected certain specimens of cancer, but nobody knew about his intentions before there were any publications. Rössle remained a famous pathologist after 1933. He managed to introduce constitutional pathology, which was attached to nationalist thinking, into the Third Reich and thus also to save the old concept of morbid anatomy (Hamperl 1972: 168–9; Schmuhl 1987; Weindling 1989).

Rössle's approach particularly influenced his student Robert Neumann (*1902), a pathologist SS physician who was linked to the concentration camps Oranienburg, Buchenwald and Auschwitz. Neumann was one of the well-sponsored scholars of Rössle, who influenced him to conduct research into twins.[38] Neumann worked at the Pathological Institute of the Charité in Berlin between 1932 and 1935, and then became head of the Pathological Institute of the Berlin-Moabit Hospital. In this capacity, he conducted experiments with concentration camp inmates at Oranienburg near Berlin, in which he used the so-called *Histotom* he had developed to collect tissue specimens of the liver from living patients. In 1939/40, he used this instrument in the Buchenwald concentration camp. A witness noticed that all of Neumann's victims died after this treatment. In Buchenwald, Neumann conducted further experiments with prisoners and performed autopsies after they had died. He also worked in the Auschwitz concentration camp. From 1940 to at least 1943, he was the head of the Pathological Institute of the German Medical Academy in Shanghai/China. In a court trial in 1948, Neumann was defended by a colleague who noted that 'it is not possible now to give an expert opinion about the real cause of death of the patients'. But Neumann needed no help, because he was able to escape to Shanghai. In 1954, Neumann was a scientific assistant at the STADA pharmaceutical company in Tübingen/Germany.[39]

Other assistants of Rössle were apparently not involved in Nazi crimes, but nevertheless supported the National Socialists directly, or at least indirectly. The Charité Institute was no stronghold of resistance against National Socialism, and neither were the pathological departments of the Berlin hospitals. They, too, shared a Janus-faced history. On the one hand, there were disadvantages in the political changes. On the other hand, there were certain advantages, especially for Robert Rössle, in respect of the organization of pathology in Berlin.

In Berlin, National Socialists were especially anxious to break any resistance and to nazify all governmental structures of the city and of public institutions. On the basis of the Law for the Restoration of the Professional Civil Service (Gesetz zur Wiederherstellung des Berufsbeamtentums) of 7 April 1933, a total of 135 scholars of the University of Berlin were expelled, the largest number in any German university.[40] This policy also affected the pathology departments of the city hospitals, because a certain number of hospital pathologists were also university lecturers. But much more important were the tough measures introduced to purge public health care in the metropolis, which cannot be described here in detail (Grell 1989, esp. 63–4; Hubenstorf/Walter 1994: 46). These measures had a tremendous effect and caused damage to those in positions of influence in pathology, because National Socialists tried to fill them with their own supporters. Nine out of fourteen changes in the occupation of hospital pathology posts were politically motivated after 1933. In thirteen out of twenty-nine appointments of hospital pathologists, National Socialists were preferred.[41] This politicization of the pathology sector was alleviated when, in January 1935, the city's public health care was put under the control of the Science Ministry. Directors, assistants and voluntary assistants of the municipal hospitals could now only be appointed with the approval of the Ministry, which expressed its delight 'that the beginning of such an important planned economy of health care was voluntarily started by the capital of the Reich'.[42]

These new developments favoured those pathologists who, owing to heavy competition, previously stood little chance of getting an appointment through the regular channels. Remarkably enough, women now succeeded in achieving leading positions in the old discipline of pathology, which had traditionally been dominated by men. In 1933, for example, Else Petri was appointed as pathologist of the Hospital Am Urban in Berlin, after her boss, Edmund Mayer, had been dismissed. She had been an active member of the Nazi party since 1932. In 1939, Petri was a member and Berlin *Gauleiter* of the National Socialist Women's Organization (NS-Frauenschaft) and of the German Women's Union (Deutsches Frauenwerk). Furthermore, she was a member of the National Socialist People's Welfare (NS-Volkswohlfahrt, NSVW) and of the National Socialist Party Physicians' League (NS-Ärztebund, NSDÄB). Petri retained her job until 1944. Of course, this was not a sign of women's emancipation, but rather a clear example of the changes caused by politically biased selection. It was even possible for hospital pathologists to receive teaching positions at the University of Berlin, based on their party membership. In 1937, for example, the pathologist Walter Benoit (*1901) applied for the transfer of his teaching licence from Jena to Berlin, where he worked as a pathologist at the Hospital Berlin-Reinickendorf. In spite of the resistance of Robert Rössle and the medical faculty, Benoit became a faculty member. Benoit had been an active member of the SA (Storm Troopers/Brown Shirts), and he was a member of several other Nazi organizations. Because of the difficulties Rössle had with the Nazi government, Reich health leader Leonardo Conti suggested Benoit as Rössle's successor as director of the Charité Institute in 1940.[43] Only a very few hospital pathologists could or wanted to avoid close connections with the Nazi regime (Prüll 1999a: 285–324). On the other hand, those who belonged to 'undesirable' social groups, such as the Jews, were removed mercilessly and occasionally with violence. During the Weimar period, the Hospital Berlin-Moabit was said to be 'red' and 'Jewish'. In April 1933, Jewish and social democratic physicians were carried off from the wards in the course of several SA campaigns. Berthold Ostertag (1895–1975), who until then had been pathologist at the Berlin Hospital Berlin-Buch, entered the Pathological Institute at the Moabit Hospital, dressed in an SA uniform, and kicked his colleague Rudolf Jaffé (1885–1975) out of his office (Pross 1993: 100–5; Pross 1984a, esp. 169–70; Pross 1984b: 189, 200).

Although these developments did not favour the discipline of pathology as such, Robert Rössle tried to keep his influence over the appointment of hospital pathologists and the acceptance of their lecturer's licences in Berlin. His policy was directed towards a centralization, not only of the discipline within his own institute, but of the discipline in the whole of Berlin. This is especially true in the case of Berlin-Moabit. First, Rössle was able to thwart National Socialist plans to create a new competitive medical faculty at this hospital. In this he was helped by several faculty members, among others his friend Ferdinand Sauerbruch (Prüll 1999a: 314–18).[44] There are serious indications that it was Rössle who forced Franz Büchner (1895–1991), at that time pathologist at the Hospital Berlin-Friedrichshain, to refuse the politically motivated offer to become pathologist at the Moabit hospital.[45] Secondly, Rössle successfully influenced all appointments to the Moabit pathology chair after 1935. His main aim was to install colleagues who were devoted to him, and to avoid competition. Any ambitions to counter Nazi policy and to uphold scientific values were only of secondary importance. This was the case when Rössle initiated the appointment of Robert Neumann at the Moabit hospital in 1935.

Another method used to control the role of hospital pathologists was to restrict their participation in the education of medical students. This was also a measure used by Rössle's predecessors, especially Lubarsch. Rössle himself intensified these efforts, especially after 1933, owing to the pressure caused by the growing number of students (Prüll 1999a: 65–81, 258–9). Although Berthold Ostertag complained heavily about him and mobilized the dean of the faculty and the National Socialist Maximinian de Crinis (1889–1945), professor of psychiatry at Berlin University, Rössle refused to accept the right of hospital pathologists to make out certificates for the students for demonstration and autopsy courses. Furthermore, he defended his right to present the main lecture course on morbid anatomy in Berlin.[46]

In summary, it can be said that Rössle successfully exploited the situation after 1933 in order to defend the traditional concept of morphological pathology and the idea of centralized organization of the discipline, which had been set up in the late nineteenth and early twentieth centuries. Although work in pathology suffered enormously during wartime, and although the institutes and pathological collections were destroyed to a great extent by the Allied bombing raids, the reinstallation of pathology in Berlin after 1945 was tied to this old concept and to old organizational patterns (Prüll 1999a: 300–24, 392–6).

Summary and Conclusion

Pathology was organized or organized itself within the context of the social, political and cultural conditions of the two cities. Pathology in Berlin, which had its origin in the German Empire, was state-financed, and organized centrally and autocratically. As a profession it was indebted to the study of morbid anatomy, which was seen as a specifically German cultural contribution to the field of pathology. Therefore, in the time of the socially and politically unstable Weimar Republic (Winkler 1993; Peukert 1987) pathology in Berlin was oriented towards pan-German ideological ideas and maintained a nationalist pathology. This meant that the organization of morbid anatomy could easily be transferred to the period after 1933, when the physician was conceptualized by the Nazis as a health leader (*Gesundheitsführer*) (Fahrenbach/Thom 1991). Rössle was able to strengthen the influence of German morbid anatomy, but above all his own authority within Berlin pathology. Those disciplines which represented laboratory medicine were abandoned. Innovative approaches, which were mostly initiated by Jewish scientists, were also rejected. The anti-Semitic remarks of Koch and Wätjen show that in this process political and scientific ideas fed off one another. Berlin pathology could thus perform its work on the basis of the traditional anti-democratic *Ordinariensystem* (professorial system). The latter was created during the German Empire and contributed to the German successes in scientific medicine. But it also preserved the power of the professor or director – at the Charité and also at the

hospital institutes. This system aided Rössle's centralizing policy and
was suitable for his plans as well as for the new 'leadership-principle'
of the Nazi period. Even today morbid anatomy plays an excessive
role in German pathology. Its declining importance since 1900 is
sometimes ignored. In 1990, for example, one representative of
morbid anatomy, Hans-Werner Altmann, wrote: 'Among us the
equation of pathology with morbid anatomy seems to remain as well
as the dissociation from laboratory medicine, not to speak about other
fields, which are tempting or important for Americans' (Altmann
1990, esp. 365).[47]

By contrast, British clinical pathology developed under the British
constitutional monarchy with a decentralized and federal organizational structure and was governed by committees. Clinical pathology
was financed privately and got its German scientific inspirations
mainly not from Virchow, but from Cohnheim. Clinical pathology was
seen as a specifically British way of working in the field, dissociating
itself from German morbid anatomy. It was performed within the walls
of the London voluntary hospitals. These were conceptualized as
philanthropic enterprises, and oriented towards traditional Victorian
values. Although because of its social policy the British government
gained more influence in the organization of health care, this situation
persisted in London up until the twentieth century. This was possible
because the work of the London voluntary hospitals corresponded to
the stable conditions in British policy and health care. Britain made
a further step towards political democratization in the inter-war
period and the party system was accepted (Johnson 1992, esp. 327–
8). Clinical pathology was able to establish itself as a profession and
Topley as well as Panton held important positions in science and
health care organizations. Pathology in London was always under the
control of external scientific committes and/or lay people. This
created problems for the maintenance of independent work. But it
guaranteed close contact with contemporary medical problems.
Competition in the private market seemed to be a stimulus in this
direction, because the interests of the clients had to be acknowledged.
For this reason, and because of its complex and federalistic structure,
clinical pathology mirrors the rules of the democratic state, which
guaranteed its creation. Even today clinical pathology in London
remains an important aspect of the discipline (see Royal College of
Pathologists 1989 and 1992). It is and was in the words of the clinical
pathologist James Henry Dible (1889–1971) 'that characteristically
English solution of a difficulty: a compromise' (Dible/Davie 1945: vi).

Notes

1. It is impossible here to quote the large number of publications dealing with medicine in the Third Reich. See as early eye-witness reports Kogon (1974) and Mitscherlich/Mielke (1997).
2. Information about the exact date of the foundation of the Pathological Institute at the Hospital at Berlin-Zehlendorf is missing.
3. Concerning London, see also Porter (1994). Porter gives a good overview and a lot of hints on further reading about all aspects of London's history. Cf. also Pennybacker (1995) and Saint (1989).
4. Conditions in Berlin pathology corresponded to German scientific organization in general. See Kater (1985, esp. 680–1), Ringer (1983) and Sontheimer (1994).
5. The Charité Administration to the Sanitary Office of the Gardecorps (Sanitätsamt des Gardekorps), Berlin, 5 January 1918, in Akten betr. die Assistenzärzte beim Pathologischen Institut 1886–1926, University Archive of the Humboldt University Berlin (in the following: UA der HUB), Charité-Direktion, Nr. 933, lf. 228; Medical Staff of the Royal Free Hospital, Meeting of the Sub-Committee for the employment of laboratory assistants, November 1916, in Royal Free Hospital. Medical Staff Minute Book 1914–1917, pp. 166–7, esp. p.167, Royal Free Hospital Medical School, Archives Centre.
6. For the domination of clinical pathology, see Gray (1918), Wigham (1928, esp. 122) and Boycott/Cameron (1916–1917, esp. 455). Concerning autopsies, 'stationary' and 'mobile' laboratories, see Meyer/Dew/Stokes (1916), Turner (1922, esp. 167), Boney/Grossman/Boulenger (1917) and Richards (1922, esp. 115, 121).
7. Gas experimentation was performed, among others, with cats to analyse the process of intoxication with phosgene. This was done from 1915 by Otto Heitzmann, army physician (*Stabsarzt*) and former member of the German colonial forces in South Africa between 1905 and 1907. He worked as an assistant of the anatomical department of the Charité Institute. Heitzmann's results were published in two papers, cf. Woll-schläger (1990: 58–9, 63, 158). For Heitzmann's career as a member of the colonial forces and as a member of the Pathological Institute of the Charité in Berlin, see: The Charité Administration to the Ministry of Science, Berlin, 19 March 1912; Curriculum vitae of Otto Max Karl Heitzmann, Düsseldorf, 14 March 1912, in Akten betr. das Pathologische Institut bei der Universität Berlin (April 1908–March 1914), Geheimes Staatsarchiv Preußischer Kulturbesitz, Berlin (including the former department in Merseburg) (in the following: GStA PK, M = Merseburg), I. HA Rep 76 Va, Kultusministerium, Tit. X, Nr. 153, Bd. III (M), lf. 312, 313.
8. Medical School Committee, Meeting, 24 November 1927, in The Middle-sex Hospital. Medical School Committee Minutes, vol. 10, September

1926–June 1933, lf. 45–7, esp. lf. 47, Leon L. Fine, University College London, Rayne Building, London. I am especially grateful to Leon L. Fine for giving me access to sources on the Middlesex Hospital, London.

9.   On behalf of the War Office, Leishman had founded a Pathological Department in 1918, and he was its first director. In 1923, he became Director General of the Army Medical Service, cf. Cummins (1926, esp. 524).

10.   Cf. Westminster Hospital Reports, vol. 6, 1927–1931, esp. 1929, 1930, in Greater London Public Record Office (in the following: GLPRO), H O2/WH/A35/02/6, pp. 26–7 (1929), p. 25 (1930).

11.   The *Journal of Clinical Pathology* was founded in 1947, see Dyke (1961: 138).

12.   E. A. Sharp, The Town Clerk, The Clerk to the Council, The Secretary or House Governor, Circular 2152 of the Ministry of Health, 26 September 1940, in Westminster Hospital, Pathological Services: Review of Staff (1 File), 1939 December–1940 September, in GLPRO, H 02/WH/A62/04, unnumbered pages.

13.   Power (Secretary, Westminster Hospital), to H. W. Vines (Sector Pathologist), London, 11 March 1940 and 19 February 1940; Vines to the Secretary (Westminster Hospital, London), 22 February 1940; Ministry of Health, Emergency Medical Service, London Emergency Pathological Service. To the Sector Pathologists for Information, no date; Fraser, Ministry of Health, to the Secretary (Westminster Hospital, London), 22 February 1940; Vines (Sector Pathologist) to the Secretary (Westminster Hospital, London), 26 February 1940; Secretary (Westminster Hospital), to Vines, London, 29 February 1940, ibid.

14.   Concerning the development of penicillin, see MacFarlane (1985: 261–74), MacFarlane (1979), Williams (1984) and L. Bickel (1972). For Fleming's biography and the investigation of penicillin with all its myths, see also Pieroth (1992), Neushul (1993: 371–95), M. H. Bickel (1994) and Hare (1970; 1982; 1983).

15.   Hubert Maitland Turnbull, unpublished autobiography, written in 1953, in The Royal London Hospital Archives and Museum, p. 53.

16.   'Massenexperiment', in which 'genotypisch ganz verschieden reagierende Individuen in einander ähnliche Formen gepresst werden' (Rössle/Böning 1927: 5).

17.   Cf. Baader (1931: 8–9); The director of the Reich's Health Office (Präsident des Reichsgesundheitsamtes) to Aschoff, Berlin, 27 November 1926; The director of the Reich's Health Office to the heads of pathological institutes in the German Reich, formal letter, Berlin, 27 November 1926; distribution list of teachers, in Obituary Ludwig Aschoff, Universitätsarchiv der Universität Freiburg, Abteilung VI/1, Wissenschaftliche Korrespondenz 1 June 1925–28 February 1927, Deutschland.

18. Report of the Reich's Health Office (Reichsgesundheitsamt) for 1927, Berlin, 3 January 1928, Report of the Reich's Health Office for 1928, Berlin, 7 January 1929, in Akten betr. Geschäftsberichte des Reichsgesundheitsamtes January 1928–March 1932, Bundesarchiv Berlin, Berlin-Lichterfelde (in the following: BArchiv Berlin), Reichsministerium des Innern, vol. 14, 1. Abteilung II, Nr. 26289, lf. 2ff., esp. lf. 68, 69 (1927); lf. 111ff., esp. lf. 218 (1928); see also Gruber (1935: 340–5, esp. 343); Busch to Aschoff, Berlin, 9 August 1929 and 11 July 1929; Aschoff to Busch, Freiburg, 17 October 1927, Busch to Aschoff, Berlin, 9 October 1927, in Obituary Aschoff, Abteilung VI/2, Wissenschaftliche Korrespondenz 1927–1929, Deutschland.

19   'die Auswirkungen des "Kampfes im Beruf" erfordern . . . von Tag zu Tag mehr unser Interesse' (Aschoff et al. 1931). See furthermore the letters of Koch to Aschoff, Berlin-Charlottenburg, 8 April 1930, Berlin-Wilmersdorf, and 28 April 1930, in Obituary Aschoff, Abteilung VI/3; Wissenschaftliche Korrespondenz 1930–1931, Deutschland.

20. 'Die Gewerbepathologie ist ja eigentlich die Kriegspathologie des Friedens' (Koch to Aschoff, Berlin-Charlottenburg, 8 April 1930, ibid.).

21. 'nachdem die deutsche Wehrmacht im Leben unseres Volkes wieder ihren alten Platz einnimmt'. See the correspondence between Ludwig Aschoff and Walter Koch, Ludwig Pick, Martin Benno Schmidt, in Obituary Aschoff, Abt.II/2(a), Zieglers Beiträge, Autoren 33–6; Vorwort der Herausgeber (esp. Ludwig Aschoff), in Welz (1936: two pages, not paginated).

22. Paul Schürmann to the Dean of the Medical Faculty, Berlin, 14 October 1935, in Personalakte Paul Schürmann (Prof. Dr. med., 12 September 1894–2 July 1941), 24 October 1934–2 June 1944 (Medizinische Fakultät), UA der HUB, Nr. S 277, lf. 8; Murken (1977: 927–33, esp. 929–32).

23. See furthermore: The Racepolicy Office of the Nazi party (*Rassenpolitisches Amt der NSDAP*), Circular No. 17, Berlin, 9 July 1934, in Fonds für Erb- und Rassenpflege May–October 1934, BArchiv Berlin, Reichsministerium des Innern, vol. 14, 1. Abteilung II, Nr. 26369, lf. 37; Ludwig Aschoff to Walter Koch, Freiburg, 14 September 1933, in Obituary Aschoff, Abteilung VI/6, Wissenschaftliche Korrespondenz 1932–1933 II: K-P (a).

24. During the Weimar period, the following pathologists worked as heads of an institution in their field in Berlin: Otto Lubarsch, Robert Rössle, Carl Hart, Walter Steinbiß, Georg Schaetz, Berthold Ostertag, Carl Benda, Rudolf Jaffé, David von Hansemann, Erwin Christeller, Hans Anders, Ludwig Pick, Karl Löwenthal, Max Busch, Ernst Walkhoff, Heinrich Ehlers, Philipp Adolf Rheindorf, Curt Froboese, Max Koch, Edmund Mayer, Max Versé, Wilhelm Ceelen, Walter Koch, Karl-Heinrich Plenge, Julius Wätjen. The following scientists had a definitely

German-nationalist attitude: Lubarsch, Jaffé, von Hansemann, Busch and Koch. Walkhoff presumably had a hostile attitude to democracy since he held NSDAP membership between April 1929 and July 1930, and from May 1933; cf. Mitgliedskartei der NSDAP, Karteikarte Ernst Walkhoff, in BArchiv Berlin, Material of the former 'Berlin Document Centre' (BDC). See furthermore, in the case of Ostertag, as an active member of the anti-republican *Freikorps* in 1919: Personalbogen Berthold Ostertag 1935, in Bundesarchiv Koblenz, Außenstelle Dahlwitz-Hoppegarten, Akten des Ministeriums für Staatssicherheit der ehem. DDR, ZW 436 A.4 (concerning Berthold Ostertag).

25. 'straffe und stramme Disziplinierung des Geistes' (Lubarsch 1917: 1380).

26. 'Aufstand der kleinen Leute, des Proletariats' (Lubarsch 1931: 591).

27. Wätjen's duty was 'den sauberen Geist Orths wahren' (Walter Koch to Ludwig Aschoff, Berlin-Wilmersdorf, 27 December 1929, in Obituary Ludwig Aschoff, Abteilung VI/2).

28. 'Grossreinemachen' (Julius Wätjen to Ludwig Aschoff, Berlin, 2 September 1929, ibid.).

29. 'völlig in Judenhänden', NSBO-Zelle der Charité to the Arbeitsausschuß, Unterabtlg. Staatl. Kliniken der Fachgruppe Gesundheitswesen: Betr. die Anfrage 'Juden und Ausländer im Patholog. Institut d. Charité', Berlin, 14 November 1932, in Akten betr. das Pathologische Institut der Universität Berlin, vol. VII., January 1932–March 1938, BArchiv Berlin, Reichsministerium für Wissenschaft, Erziehung und Volksbildung, Nr. 1406, lf. 146–7, esp. lf. 146.

30. 'sachliches Material über die sogenannten "wissenschaftlichen" Arbeiten von Juden zusammenzutragen', ibid, lf. 147; 'dass an dem ersten und grössten deutschen Institut für Pathologie deutsche Ärzte angestellt werden', Dieckmann to the NSBO Zelle of the Charité, Berlin, November 1932, ibid., lf. 148.

31. Kultusministerium, notice, 1 June 1933, ibid., lf. 149. Max H. Kuczynski, the head of the bacteriological department, was dismissed also because of the Gesetz zur Wiederherstellung des Berufsbeamtentums of 7 April 1933, but he had already left the Institute in October 1932. Cf. Hubenstorf (1994, esp. 622, 626), Roulet (1966: 501), Asen (1955: 108, 162) and Gerstengarbe (1994). See also The Charité Administration to the Ministry of Science, Berlin, 2 June 1933, BArchiv Berlin, Reichsministerium für Wissenschaft, Erziehung und Volksbildung, Nr. 1406, lf. 160–2, esp. lf. 160–1.

32. The Charité Administration to the Ministry of Science, Berlin, 11 December 1933 and 1 October 1936, ibid., lf. 189–90, 379–80. Concerning the Pathological Institute of the Charité in the Third Reich, see furthermore the very scanty and incomplete description by Udo Lampert (1991a: 76–85).

33. Robert Rössle to the Ministry of Science, Berlin, 5 November 1932, in BArchiv Berlin, Reichsministerium für Wissenschaft, Erziehung und Volksbildung, Nr. 1406, lf. 94–5, esp. lf. 94.

34. 'Fragen der Pathologie' (Rössle 1930, esp. 580). See also Robert Rössle to the Ministry of Science, Berlin, 5 November 1932, in Akten betr. die Anstellung und Besoldung der ordentlichen und außerordentlichen Professoren der Medizinischen Fakultät (July 1931–March 1933), GStA PK, I. HA Rep.76 Va, Kultusministerium, Tit IV, Nr. 46, Bd. XXVIII, lf. 421 (M); The Ministry of Science to the Charité Administration, Berlin, 5 November 1932, in Akten betr. das Pathologische Institut 1928–1932, UA der HUB, Charité-Direktion, Nr. 932, lf. 266.

35. The Ministry of Science to Robert Rössle, Berlin, 28 February 1929, in Akten betr. das Pathologische Institut bei der Universität Berlin (April 1927–December 1931), GStA PK, I. HA Rep. 76 Va, Kultusministerium, Tit. X, Nr. 153, Bd. VI, lf. 145–7, esp. lf. 145 (M); Robert Rössle to the Dean of the Medical Faculty of the University of Berlin, Basle, 16 April 1929, in Akten betr. Professoren 1928–1930, UA der HUB, Medizinische Fakultät – Dekanat, Nr. 1388, lf. 283; Charité Administration to the Ministry of Science, Berlin, 5 February 1934; Karl Hinsberg to the Ministry of Science, Berlin, 1 February 1934; Ministry of Science to the Charité Administration, Berlin, 16 March 1934, BArchiv Berlin, Reichsministerium für Wissenschaft, Erziehung und Volksbildung, Nr. 1406, lf. 210–13.

36. 'städtisches heruntergekommenes Menschenmaterial' (Rössle 1919: 34).

37. 'eine der wichtigsten Quellen der Erkenntnis vom Wesen der Krankheiten' (Rössle 1939: 284); 'therapeutisch hoffnungsloser(n) Nachwuchs(es)' (Rössle 1934: 26).

38. Ernst Leupold, Cologne, to the Reich Ministry of Science and Education (Reichsminister für Wissenschaft, Erziehung und Volksbildung), report on Robert Neumann, Cologne, 27 February 1941, in Reichsministerium für Wissenschaft, Erziehung und Volksbildung, Personalakte Robert Neumann, Allgemeine Pathologie, Berlin, Bd. I, January 1936, N 83, lf. 64, in BArchiv Berlin, ehem. Bestände des Berlin Document Center (BDC), R 2 Pers., Robert Neumann, REM 8000001026, A 47.

39. Hessian State Ministry. The Minister for Political Liberation (Hessisches Staatsministerium. Der Minister für politische Befreiung). Chamber (*Spruchkammer*) Darmstadt-Lager, to Emil Carlebach, Association of the Persecutees of the Nazi Regime (*Vereinigung der Verfolgten des Nazi-Regimes* (VVN)), Darmstadt, 31 May 1948; 'Arztschreiber Poller gegen SS-Arzt Neumann', *Der Tagesspiegel*, 23 September 1948, newspaper clipping (see the quotation here: 'man könnte heute kein Gutachten mehr über die wahre Todesursache der Patienten abgeben'); Oskar Holewa to the Hessian State Ministry. The Ministry for Political Liberation. Chamber Darmstadt-Lager, 26 April 1948; The Hessian State Ministry. The

Ministry for Political Liberation. Chamber Darmstadt-Lager, to the State Commissioner (Staatskommissar) Auerbach at the Special Ministry for Political Liberation, Munich, Darmstadt, 4 March 1948, in Bundesarchiv Koblenz, Außenstelle Dahlwitz-Hoppegarten, Akten des Ministeriums für Staatssicherheit der ehem. DDR, ZM 854 A. 16 (concerning Robert Neumann); Staff of Concentration Camps ('Besetzung der Konzentrationslager'), in B Archiv Berlin, Ordner SS, SL 19a, p. 32: See also Poller (1947: 219–23), Klee (1997: 36, 48, 55), Pross (1984b, esp. 195–6) and Prüll (1999a: 318–22).

40. List of dismissed scholars on the basis of the Law for the Reconstitution of the Professional Civil Service ('Liste: Verabschiedete oder entlassene Dozenten aufgrund des *Gesetzes zur Wiederherstellung des Berufsbeamtentums*'), in Angelegenheiten der Hochschullehrer, May 1933–June 1934 (part 2), BArchiv Berlin, Reichsministerium des Innern, Bd. 14, 2. Abteilung III, Nr. 26890/1, lf. 646.

41. The position of the pathologist at the *Urban-Krankenhaus* remained in the hands of Nazi supporters during the Third Reich. These were Else Petri (mentioned in the text) 1933–44; Gerhard Katerbau (* 1904) in 1944/5, who was also party member from 1932 and member of the N.S.-Physicians' League, of the German Labour Force (Deutsche Arbeitsfront) and the N.S.-People's Welfare. Katerbau had been – presumably under Else Petri – assistant at the Am Urban hospital since 1942. Hildegard Schönberg-Lutz was (from 1934) pathologist at the hospital Berlin-Buch, and then (since 1935) at the Berlin Kaiser und Kaiserin Friedrich Kinderkrankenhaus. Hans-Rudolph Döhnert, party member since May 1933, became successor of Ernst Walkhoff at the hospital Berlin-Lichterfelde in 1938, and he was a member of several party organizations and a leading member of the SA. Karl Heinrich Plenge was a party member and member of the SA from May 1933; in 1934, he became pathologist at the hospital Berlin-Neukölln I (Gruber mentions the year 1931, but this is wrong: Plenge's predecessor Heinrich Ehlers retired in 1934, cf. Plenge 1950, esp. 433). The National Socialist Berthold Ostertag filled the pathologist positions at the Berlin-Moabit and Rudolf Virchow hospitals in 1933 and 1934. In 1937, Walter Benoit was appointed pathologist at the Berlin-Reinickendorf hospital, Edmund Randerath in 1943 received the chair at the Berlin Hedwig's hospital, cf. Gruber (1949: 174–5) and Peiffer (1998, esp. 107). See also the index cards of the *Parteistatistische Erhebung* 1939 in BArchiv Berlin, Bestände des ehem. BDC (Berlin Document Center): Else Petri; Reichsärztekammer/NSDAP: Gerhard Katerbau, Hans Rudolph Döhnert; NSDAP/SA: Hildegard Schönberg. See also Rabl (1950: 32).

42. 'daß der Anfang einer so bedeutenden Planwirtschaft der Gesundheitspflege freiwillig von der Reichshauptstadt ausgeht', The Reich and Prussian Ministry of Science and Education (Der Reichs- und Preußische Minister für Wissenschaft, Erziehung und Volksbildung)

to the State Commissioner of Berlin, Berlin, 21 January 1935, in Akten
betreffend die Anstellung von Ärzten in den städtischen Kranken-
häusern der Reichshauptstadt Berlin, BArchiv Berlin, Reichsministerium
für Wissenschaft, Kunst und Volksbildung, Nr. 1478, lf. 2.

43. NS-Questionnaire, 22 January 1938, in Personalakte Walter Benoit, Z/
B II 1927, Akte 18, K 63 (unnumbered pages), UA der HUB; Walter
Benoit, Karteikarte des Reichserziehungsministeriums, in BArchiv
Berlin, ehem. Bestände des Berlin Document Center (BDC). For Benoit
see furthermore Prüll (1999a: 313–14).

44. Material on the Meeting of a Committee and on the Meeting of the
Medical Faculty, Berlin, no date, in Akten betr. die Heranziehung
städtischer Krankenhäuser zum Universitätsunterricht in Berlin
(*Robert-Koch-Krankenhaus*), vol. 2, July 1937–February 1944, BArchiv
Berlin, Ministerium für Wissenschaft, Erziehung und Volksbildung, Nr.
1466, lf. 9–16.

45. Report of the N.S. Lecturers' League, *Gau* Berlin, on Franz Büchner,
Berlin, 4 June 1936, in Universität Freiburg. Generalakten, 1935–1943,
Rektorat, B 1, Nr. 1222, unnumbered pages, in Universitätsarchiv der
Universität Freiburg.

46. Berthold Ostertag to the Dean of the Medical Faculty Kreuz, Berlin,
12 February 1940, in Personalakte Berthold Ostertag, (*Feb. 28, 1895,
Privatdozent Dr. med.) 6 July 1934–30 January 1946 (Medizinische
Fakultät), UA der HUB, Nr. O 49, lf. 60–3, esp. lf. 62; Berthold Ostertag
to Maximinian de Crinis, Berlin, 17 January 1941; The Reich Ministry
of Science and Education to Berthold Ostertag, Berlin, 10 March 1941,
in BArchiv, Außenstelle Dahlwitz-Hoppegarten, ZW 436 A.4 (betr.
Berthold Ostertag), lf. 48–9, 53; The Dean of the Medical Faculty to
the Reich Ministry of Science and Education, Berlin, 19 February 1941;
Berthold Ostertag to the Dean of the Medical Faculty in Rostock, Berlin,
8 May 1942, both in UA der HUB, Personalakte Nr. O 49 (betr. Berthold
Ostertag), lf. 70, 78.

47. 'Bei uns scheint es . . . bei der Gleichsetzung von Pathologie und
Pathomorphologie bleiben zu wollen und bei einer Abgrenzung von der
Labormedizin, gar nicht zu reden von den anderen Fachgebieten, die
den Amerikanern verlockend oder verpflichtend erscheinen'. Although
Altmann analyses the history of pathology since the mid-eighteenth
century critically, he argues against a 'biochemical pathology' (*biochemische Pathologie*) and calls for a new recognition of morbid anatomy,
ibid. 365–6.

References

Ackerknecht, Erwin H. (1957), *Rudolf Virchow. Arzt, Politiker, Anthro-
pologe*, Stuttgart: Enke.

—— (1967), *Medicine at the Paris Hospitals, 1794–1848*, Baltimore/MD: Hopkins.

Alter, Peter (1993), 'Einleitung', in Peter Alter (ed.), *Im Banne der Metropolen. Berlin und London in den zwanziger Jahren* (Veröffentlichungen des Deutschen Historischen Instituts London 29), Göttingen/Zurich: Vandenhoeck & Ruprecht, 7–20.

Altmann, Hans-Werner (1990), 'Die Pathologie an der Schwelle des neuen Jahrhunderts', *Würzburger Medizinhistorische Mitteilungen*, 8: 351–68.

Anselm, Sigrun (1987), 'Emanzipation und Tradition in den 20er Jahren', in Sigrun Anselm and Barbara Beck (eds.), *Triumph und Scheitern in der Metropole. Zur Rolle der Weiblichkeit in der Geschichte Berlins*, Berlin: Reimer, 253–74.

Aschoff, Ludwig (1916), 'Über die Aufgaben der Kriegspathologie', in *Kriegspathologische Tagung in Berlin am 26. und 27. April 1916* (Centralblatt für allgemeine Pathologie und pathologische Anatomie, supplement of vol. XXVII), Jena: Gustav Fischer, 1–9.

—— et al. (eds) (1920), *Veröffentlichungen aus der Kriegs- und Konstitutionspathologie*, vol. 1, Jena: Gustav Fischer.

—— (ed.) (1921), *Pathologische Anatomie* (Handbuch der Ärztlichen Erfahrungen im Weltkriege 1914/1918, ed. by Otto von Schjerning, vol. VIII), Leipzig: Barth.

—— et al. (eds) (1931), *Veröffentlichungen aus der Gewerbe- und Konstitutionspathologie*, vol. 7, issue 1, Jena: Gustav Fischer.

Asen, Johannes (ed.) (1955), *Gesamtverzeichnis des Lehrkörpers der Universität Berlin*, vol. I, 1810–1945, Leipzig: Harrassowitz.

Baader, Ernst W. (1931), *Gewerbekrankheiten*, Berlin/Vienna: Urban & Schwarzenberg.

Beckson, Karl (1992), *London in the 1890s. A Cultural History*, New York/London: Norton & Company.

Berridge, Virginia (1990), 'Health and Medicine', in F. M. L. Thompson (ed.), *The Cambridge Social History of Britain 1750–1950*, vol. 3: *Social Agencies and Institutions*, Cambridge: Cambridge University Press, 171–242.

Bickel, Lennard (1972), *Rise up to Life. A Biography of Howard Walter Florey who Gave Penicillin to the World*, New York: Charles Scribner's Sons.

Bickel, Marcel H. (1994), 'Die Anfänge der Chemotherapie: Ehrlich, Domagk, Bovet', in Günther Stille (ed.), *Der Weg der Arznei. Von der Materia Medica zur Pharmakologie*, Karlsruhe: G. Braun, 311–26.

Bleker, Johanna and Jachertz, Norbert (eds) (1993), *Medizin im 'Dritten Reich'*, Cologne: Deutscher Ärzte-Verlag.

Boney, T. K., Grossman, L. G. and Boulenger, C. L. (1917), 'Report of a Base Laboratory in Mesopotamia for 1916, with Special Reference to Water-Borne Diseases', *Journal of the Royal Army Medical Corps*, 29: 409–23.

Boycott, Arthur Edwin and Cameron, Hector (1916–17), 'Major Sydney Domville Rowland (1872–1917)', *Journal of Pathology and Bacteriology*, 21: 453–6.

Bracegirdle, Brian (1993), 'The Microscopical Tradition', in William F. Bynum and Roy Porter (eds), *Companion Encyclopedia of the History of Medicine*, vol. 1, London/New York: Routledge, 102–68.

Brocke, Bernhard vom (1990), 'Forschung und industrieller Fortschritt: Berlin als Wissenschaftszentrum. Akademie der Wissenschaften, Universität, Technische Hochschule und Kaiser-Wilhelm-Gesellschaft', in Wolfgang Ribbe and Jürgen Schmädeke (eds), *Berlin im Europa der Neuzeit. Ein Tagungsbericht* (Veröffentlichungen der Historischen Kommission zu Berlin 75), Berlin/New York: de Gruyter, 165–97.

Büsch, Otto and Haus, Wolfgang (1987), *Berlin als Hauptstadt der Weimarer Republik 1919–1933* (*Berliner Demokratie 1919–1985*, vol.1) (Veröffentlichungen der Historischen Kommission zu Berlin 70/1), Berlin: de Gruyter.

Cummins, Sydney Lyle (1926), 'William Boog Leishman 1865–1926', *The Journal of Pathology and Bacteriology*, 29: 515–27.

Cunningham, George (1992), *The History of British Pathology*, Bristol: White Tree Books.

Daunton, Martin J. (1993), 'Vorstadt, Gesellschaft und der Staat: London in den zwanziger Jahren', in Peter Alter (ed.), *Im Banne der Metropolen. Berlin und London in den zwanziger Jahren* (Veröffentlichungen des Deutschen Historischen Instituts London 29), Göttingen/Zurich: Vandenhoeck & Ruprecht, 87–110.

David, Heinz and Krietsch, Peter (1989), 'Geschichte des Pathologischen Instituts der Charité', *Charité-Annalen N.F.*, 9: 239–60.

Dible, James Henry and Davie, Thomas B. (1945), *Pathology. An Introduction to Medicine and Surgery*, London: Churchill.

Döring, Herbert (1975), *Der Weimarer Kreis. Studien zum politischen Bewußtsein verfassungstreuer Hochschullehrer in der Weimarer Republik* (Mannheimer sozialwissenschaftliche Studien 10), Meisenheim am Glan: Anton Hain.

Dyke, Sydney Campbell (1961), 'Organization of Clinical Pathology to Present Day', in William Derek Foster, *A Short History of Clinical Pathology*, Edinburgh: Livingstone, 124–42.

Engel, Michael (1994), 'Paradigmenwechsel und Exodus. Zellbiologie, Zellchemie und Biochemie in Berlin', in Wolfram Fischer et al. (eds), *Exodus von Wissenschaften aus Berlin. Fragestellungen–Ergebnisse–Desiderate. Entwicklungen vor und nach 1933* (Akademie der Wissenschaften zu Berlin, Forschungsbericht 7), Berlin/New York: de Gruyter, 296–341.

Engelhardt, Dietrich von (1985), 'Kausalität und Konditionalität in der modernen Medizin', in Heinrich Schipperges (ed.), *Pathogenese. Grundzüge und Perspektiven einer Theoretischen Pathologie*, Berlin and Heidelberg: Springer, 32–85.

Fahrenbach, Sabine and Thom, Achim (eds) (1991), *Der Arzt als 'Gesundheitsführer'. Ärztliches Wirken zwischen Ressourcenerschließung und humanitärer Hilfe im 2. Weltkrieg*, Frankfurt a. M.: Mabuse.

Fick, Rudolf (1924), 'Gedächtnisrede auf Johannes Orth', *Sitzungsberichte der Preußischen Akademie der Wissenschaften*. Öffentliche Sitzung vom 3.7.1924, Berlin: Preußische Akademie der Wissenschaften, xcii–xcix.

Fischer, Walther (1949), 'Rundschau und Ausblicke', in Walther Fischer and Georg Benno Gruber (eds), *Fünfzig Jahre Pathologie in Deutschland. Ein Gedenkbuch zum 50jährigen Bestehen der Deutschen Pathologischen Gesellschaft (1897–1947)*, Stuttgart: Thieme, 64–70.

Foster, William Derek (1961), *A Short History of Clinical Pathology*, Edinburgh: Livingstone.

—— (1983), *Pathology as a Profession in Great Britain and the Early History of the Royal College of Pathologists*, London: Royal College of Pathologists.

Gerstengarbe, Sybille (1994), 'Die erste Entlassungswelle von Hochschullehrern deutscher Hochschulen aufgrund des Gesetzes zur Wiederherstellung des Berufsbeamtentums vom 7.4.1933', *Berichte zur Wissenschaftsgeschichte*, 17: 17–39.

Giese, Willy (1963), 'Walter Koch (3.5.1880 bis 21.11.1962)', *Verhandlungen der Deutschen Gesellschaft für Pathologie*, 47: 423–6.

Gray, A. C. H. (1918), 'Mobile Laboratories', *Journal of the Royal Army Medical Corps*, 30: 257–71.

Green, David R. (1967), 'The Metropolitan Economy: Continuity and Change 1800– 1939', in Paul Thompson (ed.), *Socialists, Liberals and Labour. The Struggle for London 1885–1914*, London/Toronto: Routledge & Keegan Paul/University of Toronto Press, 8–33.

Grell, Ursula (1989), '"Gesundheit ist Pflicht". Das öffentliche Gesundheitswesen Berlins 1933–1939', in *Totgeschwiegen 1933–1945. Zur Geschichte der Wittenauer Heilstätten. Seit 1957 Karl-Bonhoeffer-*

Nervenklinik, ed. Arbeitsgruppe zur Erforschung der Geschichte der Karl-Bonhoeffer-Nervenklinik, scientific advice: Götz Aly, Berlin: Edition Hentrich, 49–76.

Gruber, Georg Benno (1935), 'Max Busch 2.XI.1886–29.VI.1934', *Verhandlungen der Deutschen Pathologischen Gesellschaft*, 28: 340–5.

—— (1949), 'Pathologisch-anatomische Krankenhaus-Prosekturen', in Walther Fischer and Georg Benno Gruber (eds), *Fünfzig Jahre Pathologie in Deutschland. Ein Gedenkbuch zum 50jährigen Bestehen der Deutschen Pathologischen Gesellschaft (1897–1947)*, Stuttgart: Thieme, 171–203.

Hamperl, Herwig (1972), *Werdegang und Lebensweg eines Pathologen*, Stuttgart/New York: Schattauer.

Hare, Ronald (1970), *The Birth of Penicillin and the Disarming of Microbes*, London: George Allen and Unwin 1970.

—— (1982), 'New Light on the History of Penicillin', *Medical History*, 26: 1–24.

—— (1983), 'The scientific Activities of Alexander Fleming, other than the Discovery of Penicillin', *Medical History*, 27: 347–72.

Harris, José (1990), 'Society and the State in Twentieth-Century Britain', in F. M. L. Thompson (ed.), *The Cambridge Social History of Britain 1750–1950*, vol.3: *Social Agencies and Institutions*, Cambridge: Cambridge University Press, 63– 117.

Harrison, Mark (1996), 'Medicine and the Management of Modern Warfare', *History of Science*, 34: 379–410.

Haupt, Heinz-Gerhard and Kocka, Jürgen (eds) (1996), *Geschichte und Vergleich. Ansätze und Ergebnisse international vergleichender Geschichtsschreibung*, Frankfurt a. M. and New York: Campus.

Honigsbaum, Frank (1979), *The Division in British Medicine. A History of the Separation of General Practice from Hospital Care 1911–1968*, New York: St. Martin's Press.

Hort, Irmgart (1987), *Die pathologischen Institute der deutschsprachigen Universitäten (1850–1914)*, Cologne: Diss. Med.

Hubenstorf, Michael (1994), 'Die 1933–1935 entlassenen Hochschullehrer der Medizin in Berlin', in Wolfram Fischer et al. (eds), *Exodus von Wissenschaften aus Berlin. Fragestellungen-Ergebnisse-Desiderate. Entwicklungen vor und nach 1933* (Akademie der Wissenschaften zu Berlin, Forschungsbericht 7), Berlin/New York: de Gruyter, 615–26.

—— and Walter, Peter Th. (1994), 'Politische Bedingungen und allgemeine Veränderungen des Berliner Wissenschaftsbetriebes 1920 bis 1950', in Wolfram Fischer et al. (eds), *Exodus von Wissenschaften aus Berlin. Fragestellungen–Ergebnisse–Desiderate. Entwicklungen vor und nach 1933* (Akademie der Wissenschaften zu Berlin, Forschungsbericht 7), Berlin and New York: de Gruyter, 5–100.

Johnson, Nevil (1992), 'Party Government und Parteienstaat: Vergleichende Überlegungen zur Rolle der politischen Parteien in Deutschland und Großbritannien', in Karl Rohe, Gustav Schmidt and Hartmut Pogge von Strandmann (eds), *Deutschland–Großbritannien–Europa. Politische Traditionen, Partnerschaft und Rivalität* (Arbeitskreis Deutsche England-Forschung 20), Bochum: Universitätsverlag Dr. N. Brockmeyer, 323–46.

Jünger, Ernst (1982), *Der Arbeiter. Herrschaft und Gestalt* (Hamburg 1932), Stuttgart: Klett.

Kater, Michael H. (1985), 'Professionalization and Socialization of Physicians in Wilhelmine and Weimar Germany', *Journal of Contemporary History*, 20: 677–701.

—— (1989), *Doctors under Hitler*, Chapel Hill/London: The University of North Carolina Press.

Klasen, Eva-Maria (1984), *Die Diskussion über eine 'Krise' der Medizin in Deutschland zwischen 1925 und 1935*, Mainz: Med. Diss.

Klee, Ernst (1997), *Auschwitz, die NS-Medizin und ihre Opfer*, Frankfurt a. M.: S. Fischer.

Kogon, Eugen (1974), *Der SS-Staat. Das System der deutschen Konzentrationslager* (1945) München: Kindler.

Lampert, Udo (1991a), *Die Pathologische Anatomie in der Zeit des Nationalsozialismus unter besonderer Beachtung der Rolle einiger bedeutender Fachvertreter an deutschen Universitäten und Hochschulen*, Leipzig: Medical Faculty, Thesis.

—— (1991b), 'Zur Situation der Pathologischen Anatomie an den deutschen Hochschulen während des Zweiten Weltkrieges', in Sabine Fahrenbach and Achim Thom (eds), *Der Arzt als "Gesundheitsführer". Ärztliches Wirken zwischen Ressourcenerschließung und humanitärer Hilfe im 2. Weltkrieg*, Frankfurt a. M.: Mabuse, 143–50.

Lawrence, Susan (1993), 'Medical Education', in William F. Bynum and Roy Porter (eds), *Companion Encyclopedia of the History of Medicine*, vol. 2, London/New York: Routledge, 1151–79.

Lehnert, Detlef (1993), 'Fragmentierte Gesellschaft und moderne Massenbewegungen. Zur politischen Kultur der Berliner Republikzeit', in Peter Alter (ed.), *Im Banne der Metropolen. Berlin und London in den zwanziger Jahren* (Veröffentlichungen des Deutschen Historischen Instituts London 29), Göttingen/Zurich: Vandenhoeck & Ruprecht, 196–218.

Leishman, William Boog (1923), 'Organization of the Pathological Service', in William Boog Leishman and Sir William Grant MacPherson (eds), *Medical Services Pathology* (History of the Great War. Based on Official Documents), London: Heinemann, 1–31.

Lubarsch, Otto (1916), *Wissenschaft und Volkstum*. Rede zur Feier des Geburtstages Seiner Majestät des Deutschen Kaisers König von Preußen Wilhelm II. in der Aula der Königlichen Christian-Albrechts-Universität am 27. Januar 1916, Kiel: Lipsius & Tischer.

—— (1917), 'Ueber Aufgaben und Ziele der pathologischen Forschung und Lehre', *Deutsche medizinische Wochenschrift*, 43 (no. 44): 1377–80.

—— (1927), 'Eröffnungsansprache', *Verhandlungen der Deutschen Pathologischen Gesellschaft*, 22: 1–5.

—— (1931), *Ein bewegtes Gelehrtenleben. Erinnerungen und Erlebnisse. Kämpfe und Gedanken*, Berlin: Springer.

MacFarlane, Gwyn (1979), *Howard Florey. The Making of a Great Scientist*, Oxford: Oxford University Press.

—— (1985), *Alexander Fleming. The Man and the Myth*, Oxford/New York: Oxford University Press.

Maulitz, Russell C. (1987), *Morbid Appearances. The Anatomy of Pathology in the Early Nineteenth Century*, Cambridge: Cambridge University Press.

—— (1993), 'The Pathological Tradition', in William F. Bynum and Roy Porter (eds), *Companion Encyclopedia of the History of Medicine*, vol. 1, London/New York: Routledge, 169–91.

McIntosh, James (1944), 'Experiences of a London Pathologist in Wartime', *The Middlesex Hospital Journal*, 44: 61–2.

Meyer, Martin (1993), *Ernst Jünger*, Munich: dtv.

Meyer, W. C. B., Dew, J. W. and Stokes, A. (1916), 'A Note on Four Post-mortem Examinations in which Rupture of the Intestine was found, although the Course of the Projectile was Extraperitoneal', *Journal of the Royal Army Medical Corps*, 26: 100–2.

Middlemas, Keith (1993), 'Ein Vertrag auf Dauer. Britische Politik und Öffentlichkeit in den zwanziger Jahren', in Peter Alter (ed.), *Im Banne der Metropolen. Berlin und London in den zwanziger Jahren* (Veröffentlichungen des Deutschen Historischen Instituts London 29), Göttingen/Zurich: Vandenhoeck & Ruprecht, 179–95.

Mitscherlich, Alexander and Mielke, Fred (eds) (1997), *Medizin ohne Menschlichkeit. Dokumente des Nürnberger Ärzteprozesses* (1948), Frankfurt a. M.: S. Fischer.

Moraw, Peter (1988), 'Vom Lebensweg des deutschen Professors', *Forschung. Mitteilungen der DFG*, 4: 1–12.

Moscucci, Ornella (1993), *The Science of Woman. Gynaecology and Gender in England, 1800–1929*, Cambridge: Cambridge University Press.

Murken, Axel Hinrich (1977), 'Leben und Werk des Tuberkulose-forschers Paul Schürmann (1895–1941)', *Gütersloher Beiträge*, 46: 927–33.

Neushul, Peter (1993), 'Science, Government, and the Mass Production of Penicillin', *Journal of the History of Medicine and Allied Sciences*, 48: 371–95.

Orth, Johannes (1904), *Die Stellung der Pathologischen Anatomie in der Medizin und der Pathologisch-anatomische Unterricht* (Festrede, gehalten zur Feier des Stiftungsfestes der Kaiser Wilhelms-Akademie für das militärärztliche Bildungswesen am 2. Dezember 1904), Berlin: Otto Lange.

—— (1910), 'Das Pathologische Institut', in Max Lenz (ed.), *Geschichte der Königlichen Friedrich-Wilhelms-Universität zu Berlin*, vol.3, *Wissenschaftliche Anstalten. Spruchkollegium. Statistik*, Halle: Verlag der Buchhandlung des Waisenhauses, 165–76.

Pantel, Johannes and Bauer, Axel (1990), 'Die Institutionalisierung der Pathologischen Anatomie im 19. Jahrhundert an den Universitäten Deutschlands, der deutschen Schweiz und Österreichs', *Gesnerus*, 47: 303–28.

Panton, Philip Noel (1951), *Leaves from a Doctor's Life*, London/Toronto: Heinemann.

Payer, Lynn (1989), *Medicine & Culture: Notions of Health and Sickness in Britain, the US, France and West Germany*, London: Gollancz.

Peiffer, Jürgen (1998), 'Die Vertreibung deutscher Neuropathologen 1933–1939', *Der Nervenarzt*, 69: 99–109.

Pennybacker, Susan D. (1995), *A Vision for London 1889–1914. Labour, Everyday Life and the LCC Experiment*, London/New York: Routledge.

Peukert, Detlev (1987), *Die Weimarer Republik*, Frankfurt a. M.: Suhrkamp.

Pickstone, John V. (1992), 'Introduction', in John V. Pickstone (ed.), *Medical Innovations in Historical Perspective*, Houndmills, Basingstoke and London: Macmillan, 1–16.

Pieroth, Ingrid (1992), *Penicillinherstellung: Von den Anfängen bis zur Großproduktion*, Stuttgart: Wissenschaftliche Verlagsgesellschaft.

Plenge, Karl (1950), 'Heinrich Ehlers (1875–1940)', *Verhandlungen der Deutschen Gesellschaft für Pathologie*, 32: 433–4.

Poller, Walter (1947), *Arztschreiber in Buchenwald. Bericht des Häftlings 996 aus Block 39*, Hamburg: Phönix-Verlag.

Porter, Roy (1994), *London. A Social History*, London: Hamish Hamilton.

Pross, Christian (1984a), 'Das Krankenhaus Moabit 1920, 1933, 1945. Die Jüdischen Ärzte am Krankenhaus Moabit', in Christian Pross and Rolf Winau (eds), *Nicht mißhandeln!* (Stätten der Geschichte Berlins 5), Berlin: Hentrich, 109–79.

—— (1984b), 'Die "Machtergreifung" am Krankenhaus', in Christian Pross and Rolf Winau (eds), *Nicht mißhandeln!* (Stätten der Geschichte Berlins 5), Berlin: Hentrich, 180–205.

—— (1993), 'Die "Machtergreifung" am Krankenhaus', in Johanna Bleker and Norbert Jachertz (eds), *Medizin im 'Dritten Reich'*, Cologne: Deutscher Ärzte-Verlag, 97–108.

Prüll, Cay-Rüdiger (1996), 'Die Sektion als letzter Dienst am Vaterland. Die deutsche "Kriegspathologie" im Ersten Weltkrieg', in Wolfgang U. Eckart and Christoph Gradmann (eds), *Die Medizin und der Erste Weltkrieg* (Neuere Medizin- und Wissenschaftsgeschichte. Quellen und Studien 3), Pfaffenweiler: Centaurus, 155–82.

—— (1997a), 'Otto Lubarsch (1860–1933) und die Pathologie an der Berliner Charité von 1917 bis 1928. Vom Trauma der Kriegsniederlage zum Alltag eines deutschnationalen Hochschullehrers in der Weimarer Republik', *Sudhoffs Archiv*, 81: 193–210.

—— (1997b), 'Pathologie und Politik – Ludwig Aschoff (1866–1942) und Deutschlands Weg ins Dritte Reich', *History and Philosophy of Life Sciences*, 19: 331–68.

—— (1998), 'Holism and German Pathology (1914–1933)', in George Weisz and Christopher Lawrence (eds), *Greater than the Parts. Holism in Biomedicine 1920–1950*, Oxford: Oxford University Press, 46–67.

—— (1999a), 'Medizin am Toten oder am Lebenden? – Pathologie in Berlin und in London 1900 bis 1945', typewritten Ms, Habilitationthesis Universität Freiburg (to be published).

—— (1999b), 'Pathology at War 1914–1918 – Germany and Britain in Comparison', in Roger Cooter, Mark Harrison and Steve Sturdy (eds), *Medicine and Modern Warfare* (Wellcome Institute for the History of Medicine Series), Amsterdam, Atlanta, GA: Rodopi, 131–61.

Rabl, Rudolf (1950), 'Hildegard Schönberg-Lutz (25.5.1903 bis 10.6.1939)', *Verhandlungen der Deutschen Gesellschaft für Pathologie*, 34: 32.

Ribbe, Wolfgang (ed.) (1987), *Geschichte Berlins, vol. 2, Von der Märzrevolution bis zur Gegenwart*, Munich: C. H. Beck.

—— (1990), 'Berlin im Europa der Neuzeit: Nationale Hauptstadt und europäische Metropole', in Wolfgang Ribbe and Jürgen

Schmädeke (eds), *Berlin im Europa der Neuzeit. Ein Tagungsbericht* (Veröffentlichungen der Historischen Kommission zu Berlin 75), Berlin/New York: de Gruyter, 17–51.

Richards, O. (1922), 'The Development of Casuality Clearing Stations', *Guy's Hospital Reports* (War Memorial Number), 70: 115–23.

Ringer, Fritz K. (1983), *Die Gelehrten. Der Niedergang der deutschen Mandarine 1890–1933* (Cambridge/Mass. 1969), Munich: dtv.

Rössle, Robert (1919), 'Allgemeine Pathologie und Pathologische Anatomie. Bedeutung und Ergebnisse der Kriegspathologie', *Jahreskurse für ärztliche Fortbildung*, Munich: Lehmanns, 15–46.

—— (1930), 'Aus der Geschichte des Berliner Pathologischen Instituts und der Deutschen Pathologischen Gesellschaft', *Klinische Wochenschrift*, 9: 577–82.

—— (1934), 'Zur Kritik des Konstitutionsbegriffs', in Walter Jaensch (ed.), *Konstitutions-und Erbbiologie in der Praxis der Medizin*, Leipzig: Barth, 16–26.

—— (1939), 'Erbpathologie des Menschen', in *Relazioni del IV Congresso internazionale di Patologia Comparata, Roma, 15–20 Maggio 1939*, Milano: Istituto Sieroterapico Milanese, 265–87.

—— (1940), *Die pathologische Anatomie der Familie*, Berlin: Springer.

—— and Böning, Herta (1927), 'Das Wachstum der Schulkinder. Ein Beitrag zur pathologischen Physiologie des Wachstums' (first publication: Jena 1924), *Veröffentlichungen aus der Kriegs- und Konstitutionspathologie*, issue 15, vol. 4: 1–72.

Roulet, Frederic (1966), 'Arnold F. Strauss (1902 bis 6.11.1965)', *Verhandlungen der Deutschen Gesellschaft für Pathologie*, 50: 501.

Royal College of Pathologists, London (ed.) (1989), *Codes of Practice for Pathology Departments*, London: Royal College of Pathologists.

—— (ed.) (1992), *Medical and Scientific Staffing of National Health Service Pathology Departments*, London: Royal College of Pathologists.

Saint, Andrew (ed.) (1989), *Politics and the People of London. The London County Council 1889–1965*, London/Ronceverte.

Sauerteig, Lutz (1998), 'Vergleich: Ein Königsweg auch für die Medizingeschichte? Methodologische Fragen komparativen Forschens', in Norbert Paul and Thomas Schlich (eds), *Medizingeschichte: Aufgaben, Probleme, Perspektiven*, Frankfurt/New York: Campus, 266–91.

Schmiedebach, Heinz-Peter (1993), 'Pathologie bei Virchow und Traube. Experimentalstrategien in unterschiedlichem Kontext', in Hans-Jörg Rheinberger and Michael Hagner (eds), *Die Experimentalisierung des Lebens. Experimentalsysteme in den biologischen Wissenschaften 1850/1950*, Berlin: Akademie Verlag, 116–34.

Schmuhl, Hans-Walter (1987), *Rassenhygiene, Nationalsozialismus, Euthanasie. Von der Verhütung zur Vernichtung 'lebensunwerten Lebens',* *1890–1945* (Kritische Studien zur Geschichtswissenschaft 75), Göttingen: Vandenhoeck & Ruprecht.

Sellin, Volker (1978), 'Politik', in *Geschichtliche Grundbegriffe. Historisches Lexikon zur politisch-sozialen Sprache in Deutschland*, vol. 4, Stuttgart: Klett-Cotta, 789–874.

Sontheimer, Kurt (1994), *Antidemokratisches Denken in der Weimarer Republik. Die politischen Ideen des deutschen Nationalismus zwischen 1918 und 1933* (Munich 1962) München: dtv.

Steinbach, Peter (1990), 'Berlin unter dem Nationalsozialismus', in Wolfgang Ribbe and Jürgen Schmädeke (eds), *Berlin im Europa der Neuzeit. Ein Tagungsbericht* (Veröffentlichungen der Historischen Kommission zu Berlin 75), Berlin/New York: de Gruyter, 315–28.

Stichweh, Rudolf (1994), *Wissenschaft, Universität, Professionen. Soziologische Analysen*, Frankfurt a. M.: Suhrkamp.

Thompson, Paul (ed.) (1967), *Socialists, Liberals and Labour. The Struggle for London 1885–1914*, London: Routledge & Paul.

Tröhler, Ulrich (1993), 'Surgery (modern)', in William F. Bynum and Roy Porter (eds), *Companion Encyclopedia of the History of Medicine*, vol. 2, London/New York: Routledge, 984–1028.

Turner, P. (1922), 'Some Experiences of the Work of General Hospitals in France', *Guy's Hospital Reports* (War Memorial Number), 70: 157–82.

Webster, Charles (1988), *The Health Services since the War*, vol. I, *Problems of Health Care. The National Health Service before 1957*, London: HMSO.

—— (1996), *The Health Services since the War*, vol. II, *Government and Health Care. The National Health Service 1958–1979*, London: HMSO.

Weindling, Paul (1989), *Health, Race and German Politics between National Unification and Nazism, 1870–1945*, Cambridge: Cambridge University Press.

—— (1993), 'Das goldene Zeitalter des städtischen Gesundheitswesens? Gesundheitspolitik im Berlin und London der zwanziger Jahre', in Peter Alter (ed.), *Im Banne der Metropolen. Berlin und London in den zwanziger Jahren* (Veröffentlichungen des Deutschen Historischen Instituts London 29), Göttingen/Zurich: Vandenhoeck & Ruprecht, 219–33.

Weisz, George and Lawrence, Christopher (eds) (1998), *Greater than the Parts. Holism in Biomedicine 1920–1950*, Oxford: Oxford University Press.

Welz, Alfred (1936), 'Renaler Zwergwuchs', *Veröffentlichungen aus der Konstitutions- und Wehrpathologie*, issue 38, vol.9/ issue 1, Jena: Gustav Fischer.

Wigham, J. T. (1928), 'Adrian Stokes (1887–1927)', *Journal of Pathology and Bacteriology*, 31 (I): 121–5.

Williams, Trevor I. (1984), *Howard Florey: Penicillin and After*, Oxford: Oxford University Press.

Windeyer, B. W. (1945), 'The Middlesex Hospital, 1939–1945', *The Middlesex Hospital Journal*, 45: 61–7.

Winkler, Heinrich August (1993), *Weimar 1918–1933. Die Geschichte der ersten deutschen Demokratie*, Munich: C. H. Beck.

Wollschläger, Peter (1990), *Kampfstofforschung und ärztliches Gewissen. Der Beitrag der Medizin zur chemischen Kriegführung 1914–1933*, Medical Faculty Berlin, Thesis.

Young, Ken and Garside, Patricia L. (1982), *Metropolitan London. Politics and Urban Change 1837–1981* (Studies in Urban History 6), London: Edward Arnold.

STEFAN KÜHL

The Relationship between Eugenics and the so-called 'Euthanasia Action' in Nazi Germany: A Eugenically Motivated Peace Policy and the Killing of the Mentally Handicapped during the Second World War

In the early stages of research into the so called 'euthanasia action' in Nazi Germany, historians have traced the clinical
killing of the mentally handicapped back to a eugenic ideology. The
argument was made that the killing of so-called 'lives not worth living'
(*lebensunwertes Leben*) was the 'ultimate form' of negative eugenics
(Schmuhl 1987; see also Klee 1983; Friedlander 1995: 21, 127).
However, in rejection of this theory, some scholars have argued that
the killing of the handicapped had no 'systemic place' in the ideology
of race hygienists and eugenicists (Weingart/Kroll/Bayertz 1988: 524;
Reyer 1991: 115; Schwartz 1996: 614). Neither from the logic of
'selection' nor out of fear of 'degeneration', their argument ran, could
the killing of human beings have been justified by a eugenic ideology.

A closer look at the sources supports this second view. Before 1939,
the majority of eugenicists and race hygienists did not support the
systematic killing of the mentally handicapped. They did not foresee
any positive racial improvement in the elimination of handicapped
people. They believed that there were more effective means of

preventing their reproduction. In Germany socialist, liberal and Catholic eugenicists in particular had argued against the killing of the mentally handicapped when, after the First World War, the professor of law Karl Binding and the psychiatrist Alfred Hoche launched a campaign for the killing of the so-called 'lives not worth living' (for an overview see Schwartz 1998). In fact even the great majority of leading right-wing eugenicists and race hygienists drew a clear line between eugenic measures like sterilization and marriage prohibition and the killing of the handicapped. For example in 1913, the Permanent International Eugenics Committee rejected the idea of killing the handicapped. In a programme developed by the Norwegian race hygienist Alfred Mjöen (1914: 140; for more details see Kühl 1997: 36) the international organization claimed that there was a fundamental difference between the right to live and the right to give life. While the first was a fundamental human right, the second should be a privilege only for selected, 'genetically suitable' couples.

By accepting this clear-cut distinction between laws concerning reproduction and the actual killing of the mentally handicapped in the former discourse of eugenicists, historians have to explain the actual behaviour of race hygienists and eugenicists when faced with the killing of mentally handicapped people in Germany. Leading members of the German race hygienist movement did participate in the so called 'T4 killing action', or at least they accepted the bureaucratized killings without protest (see Müller-Hill 1984; Friedlander 1995: 128; Kühl 1997: 165 and especially Massin 1996: 816–17). For example, Fritz Lenz, the first professor for race hygiene in Germany, participated in 1940 in the attempt to legalize the killing of handicapped people. He was a member of the committee that drafted a 'euthanasia law'. Ernst Rüdin, director of the Kaiser Wilhelm Institute of Psychiatry in Munich and long-time president of the International Federation of Eugenic Organizations, collaborated with leading figures of the euthanasia action to redefine the role of psychiatry in Germany. In 1942 he declared his agreement in principle with the killing of the mentally handicapped. Kurt Pohlisch, professor of psychiatry in Bonn and one of the German members of the International Federation of Eugenic Organizations, was one of the medical advisers who during the mass murder operations decided which of the handicapped would be killed. Werner Villinger, one of the leading eugenicists within the scientific community of psychiatrists, was another of the medical experts for adult euthanasia.

In the complex decision-making processes which led to the organized killing programme in Germany the behaviour of the race hygienists needs particular consideration.[1] Why did they accept drastic measures which they had rejected earlier? What was the qualitative turning point from 'hereditary and racial welfare' to the systematic extermination programmes? In this essay I shall argue that it required a profound sense of disappointment amongst race hygienists about Germany's unsuccessful peace policy to secure their acceptance and support for the comprehensive killing programmes. I want to show that the 'destruction' of their vision of a stable and peaceful state of 'superior' human beings laid them open to proposals for drastic measures which hitherto did not have a systemic place in their programme. To reconstruct this vision of an 'international eugenic peace order' I must focus on the place of eugenicists and race hygienists in international politics. The historiography of eugenics has traditionally concentrated on their attitudes towards questions of domestic policy. Their commitment to subsidies for 'genetically valuable' couples (positive eugenics) and to the prevention of the reproduction of handicapped people through marriage restrictions, sterilization and imprisonment in asylums (negative eugenics) have been extensively described by historians. Without doubt, eugenicists had their greatest influence on these matters. Their professional background in psychiatry, medicine, genetics, anthropology or population science made them experts in 'solving' the social problems of poverty, alcoholism, mental illness, criminality and prostitution. But their focus was not at all limited to domestic policy. Eugenics and race hygiene were comprehensive ideologies, claiming to provide solutions for every question facing mankind. In the first half of the twentieth century, eugenicists in different countries developed proposals for resolving problems of international relations. This vision became more and more the result of the interaction among eugenicists from different nationalities. The eugenically motivated peace policy developing in the 1910s and 1920s shows how eugenicists in Great Britain, the United States, France, Italy, Germany, Austria, Norway and Sweden were linked by their common worry about the contra-selective effects of the First World War.

First, I shall demonstrate how the attitude of eugenicists towards war shifted from a positive attitude to a much more critical position. Secondly, I shall show that the experiences of the First World War shaped their perception of war as highly 'dysgenic' and stimulated extensive discussion among them. Thirdly, I shall show how the question of war and peace was linked to the question of the mentally handicapped. Fourthly, I want to point out how the informal international contacts amongst eugenicists became more and more institutionalized. Lastly, I shall show that the Nazi government in Germany took up this international debate and used it to present their race policy as an effective peace policy. I shall try to demonstrate how this eugenically motivated peace vision could ultimately lead to an acceptance of the killing of mentally handicapped people during the Second World War.[2]

War and the 'Struggle for Survival'

At the turn of the century many scientists involved in the developing eugenics movement tended to see war as an effective means for selecting the superior qualities of a race. Adopting Darwin's concept of the struggle for existence they stressed the positive influence war had in the selection process. In Germany anthropologists and biologists like Otto Ammon and Heinrich Ernst Ziegler believed in the healthy, hygienically positive effects of war. They assumed that the struggle for survival in war was to prevent social and moral degeneration (Ammon 1895; Ziegler 1893; see also Kröner 1980: 45; Weindling 1989: 99). In Great Britain it was mostly the biometrician Karl Pearson who propagated war as an effective means for race improvement. In November 1900, at the climax of the Boer War, Pearson claimed that the struggle for existence meant suffering, but that this was the mechanism of all progress: 'This dependence of progress on the survival of the fittest race, terribly black as it may seem to some of you, gives the struggle for existence its redeeming features; it is the fiery crucible out of which comes the finer metal' (Pearson 1905: 26–7; see also Semmel 1958). Very much in line with the militarist thinking of the time, he claimed that if wars ceased 'mankind would no longer progress'. There would be nothing to check the fertility of 'inferior stock'. The relentless 'law of heredity 'would no longer be controlled and guided by natural selection. As the British historian Geoffrey Searle has pointed out, Pearson's conviction led to a rapprochement between certain British eugenicists and militarists campaigning for compulsory military service. Colonel Melville, professor of hygiene at the Royal Army Medical College, stated that military service would be eugenically useful because it inculcated in men the ideals of physical fitness, efficiency, courage and patriotism.

He argued that an 'occasional war' might be of service because in times of danger 'the nation looks to the virility of its citizens' (Melville 1910/11: 54; see also Searle 1976: 37). In the United States it was Roland Campbell Macfie who claimed in several newspapers that wars had a eugenically positive effect on the stock. He took the view that the principal eugenic consequence of wars would be a 'shortage of men' and therefore a more 'careful weeding of women' was necessary: 'War means . . . not so much a martial selection of men by blind bullets and impartial bombs as a deliberate stringent matrimonial selection of women by the critical eyes of men' (Macfie 1917a: 442). Ultimately war would lead to an improvement in the 'health and beauty of the combatant races' (Macfie 1917a: 442; see also Macfie 1917b). The thinking of Ammon and Ziegler in Germany, of Pearson and Melville in the United Kingdom, and of Macfie in the United States was the blatant application of Darwin's concept of the survival of the fittest to international relations. It fitted well into the militarist, imperialist thinking of the period and linked eugenicists with nationalist move-ments within the different countries.

Yet by at the beginning of the century other eugenicists started to think differently. Especially in the United States eugenicists were eager to stress the 'contra-selective' or – to use the technical term of eugenicists – 'dysgenic' effects of war. Vernon Kellogg, a leading American eugenicist and founder of the famous cornflakes, attacked the assumption of eugenic militarists that war's high mortality was a proof of war's benefice to the race. He claimed that, on the contrary, 'military selection is as far as possible removed from natural selection'. In his view, war was 'peculiarly unnatural' (Kellogg 1913: 102–6):

I simply cannot see the eugenic advantages of war. On the contrary, not only do I think I can see from the standpoint of the biologist and student of heredity a plausible, logical case for the dysgenic effect of war and military service, but I also believe that we have accessible, actual statistical proof of the deplorable effect.

Like Kellogg, David Starr Jordan (1910: 95; see also Jordan 1915), president of Stanford University, feared the 'inevitable impoverish-ment of the stock' by the effects of the war. The 'strongest and best men' would be the ones who were killed or injured and who would leave few or no children. The 'weaklings alive' would stay at home and beget children. Jordan and Kellogg were supported by British eugenicists like Edgar Schuster (1912: 231) from University College

London and Dean William R. Inge (1913/14) of St. Paul's Cathedral.
The stronghold of a eugenic-minded peace policy, however, was
without doubt in the United States. Irving Fisher, professor of political
economy at Yale University, warned of the 'waste of germ plasm' (see
Haller 1963: 88). Along with other eugenicists like Edward A. Ross
and Andrew Carnegie, he supported an initiative of Frank Smith, a
member of the House of Representatives, for a 'eugenic peace'. Smith
demanded cooperation between Britain, France, Germany and the
United States to ensure 'the spread of the superior human elements'.
The 'omnipotent Anglo-French-German-American-League of Civilization' would in his opinion be the 'royal road to disarmament' (Smith
1914: 2–3; see also Lenz 1914/15).

Eugenicists who saw peace as a necessary condition for improving
the racial stock did not automatically reject war as 'dysgenic'. Alfred
Ploetz, who was one of the first German race hygienists pointing to
the contra-selective effects of war, also proposed in 1895 that the 'worst
individuals' should be drafted for military service. In the case of war
'especially bad specimens' should be used as 'cannon fodder' (Ploetz
1895: 147; see also Lutzhöft 1971: 335). Paul Popenoe and Roswell
Johnson (1933: 210), authors of the main eugenic textbook in the
United States, claimed that theoretically it would be possible to
reform the process of war so that it would be mainly eugenic in effect.
This 'eugenic war' would be fought with 'elderly men as officers and
with mental defectives in the ranks'. And even Kellogg (1914: 48),
main promoter of a eugenically motivated peace policy, admitted that
military selection might be of biological advantage if it were the whole
population that was exposed.

Many eugenicists demanding a eugenic peace order agreed that
there was something like a biologically determined tendency in human
beings to fight wars. Fritz Lenz (1923: 51–3) wrote that 'most people
have a belligerent instinct'. Albert E. Wiggam (1923: 218), an
American writer and popularizer of the idea of eugenic peace, stated
that human beings naturally wanted, like animals, war, and that there
was no peace in nature. Along with this assumption about the 'nature'
of human beings, eugenicists claimed in general that the war between
'primitive tribes' had to this day a positive selective effect. Lenz (1923:
53), for example, stated that war between 'primitive people' led to
the expansion of the more capable group. Furthermore, within this
superior group the men fittest for active service would in general have
more children than the weaker men. A similar argument is used by
the British biologist J. Arthur Thomson:

In ancient days a battle was probably in many cases a sifting out of the less strong, the less nimble, the less courageous on both sides, and the result of a war or raid was probably, in some cases, the practical elimination of the weaker of two clans. In both these ways there may have been a eugenic selection of the types best suited for times when fighting was the order of the day. (Thomson 1915: 1)

Despite this transfiguration of war between tribes, it was modern warfare which many eugenicists saw as representing a great danger for their races. For example, Thomson (1915: 1) claimed that since the early struggles between clans 'times have changed and war with them'. Winning nations would no longer completely exterminate the other nations and victory would not necessarily lie with those of better physique. In his opinion, in modern wars elimination was either indiscriminate or in the 'wrong direction'. The 'finest companies' were ordered to undertake the most hazardous tasks and the 'conspicuously brave' were 'particularly liable to be killed'. For Lenz (1923: 52), war only became dysgenic with modern warfare. The defeated people would no longer be eliminated, but could continue to procreate. He gave the example of the 'race of the negroes' who could continue to have children in the United States despite their enslavement.

Eugenicists located the 'dysgenic' danger of modern war not only in the killing of 'superior stock' on the battlefields, but also in the diseases menacing the troops. Kellogg (1912: 228) pointed out that in times of war 'disease has always reaped a far greater harvest of deaths' in the army than have the bullets and bayonets of battle. In an article entitled 'The Bionomics of War' he included evidence drawn from different armies. In the twenty-year stretch of the Napoleonic campaigns six times more soldiers of the British army had been killed by diseases than by gunfire. The British losses in the two and a half years of war in the Crimea had been 3 per cent by gunfire and 20 per cent by diseases. But even in peacetime diseases would spread more widely among soldiers than among civilians. As Kellogg pointed out, in the middle of the nineteenth century the mortality rate among the armies of France, Prussia and England during times of peace was 50 per cent higher than among the civilian population. In the British army in India alone, admissions to hospital for venereal disease reached in 1895 a figure of 537 per 1,000 men (Kellogg 1914: 49–50).

In their debates with militarists those eugenicists who propagated peace as a necessary condition for race improvement referred to 'historical evidence'. Jordan stated that Rome fell only because the old Roman stock was for the most part banished or exterminated through wars: 'The Romans were gone and that was the end of it; while the sons of slaves, camp-followers, scullions, and peddlers filled the eternal city' (Jordan 1913: 140). From the perspective of Jordan and Kellogg, Napoleon's difficulties in the later years of the Wars of the Empire paralleled the earlier Roman conditions. In order to make his conscription net gather its necessary load of men, he first had to reduce, in 1799, the minimum height of conscripts fit for service from 1,624 mm to 1,598 mm. In 1804 he lowered it to 1,544 mm (Kellogg 1912: 226). Kellogg concluded:

> The actual results in racial modification due to the removal from the breeding population of France of its able-bodied male youth, leaving its feeble-bodied youth and senescent maturity at home to be the father of the new generation, is plainly visible in the condition of the conscript of later years. From the recruiting statistics, as officially recorded, it may be stated with confidence that the average height of the men of France began notably to decrease with the coming of age in 1813 and on, of the young men born in the years of the Revolutionary Wars, and that it continued to decrease in the following years with the coming of age of youths born during the Wars of the Empire. (Kellogg 1914: 46–7)

Kellogg stated that the average height of the annual conscription contingent born during the Napoleonic Wars was about 1,625 mm in size, and increased only with those born after the war. Other examples, more or less underlined by scientific data, were the decline of the Spanish Empire during the seventeenth century, the dysgenic effects of the Civil War in the United States, and the 'inferior' German and French babies born during the war of 1870–71 (Jordan 1910: 102; Jordan 1913: 140).

The discussions among eugenicists at the beginning of the twentieth century about the dysgenic or eugenic effects of war were highly controversial. At the first International Congress for Eugenics in 1912, eugenicists from different nationalities discussed under the pressure of the tense international situation the 'factors which make for racial improvement or decay' (Eugenics Education Society n.d.: 4). In the section entitled 'Sociology and Eugenics', Vernon Kellogg presented his thesis that modern war was dysgenic and had to be prevented:

> The whole army is a group of individuals not chosen at random from the population, representing both sexes, all ages, and weak and strong alike, but is already, by the very conditions of its organization, a part of the

population selected first for sex and then for ripe youth, full stature and
strength, and freedom from infirmity and disease. (Kellogg 1912: 223)

His speech was attacked particularly by eugenicists with personal or
professional links to the military. A German general claimed that
'military service is not injurious to the body but healthy, and not
depressing to mind and spirit but inspiring' (quoted according to
Kellogg 1913: 108). Arnold White, representing the British National
Service League, drew attention to the 'eugenic effect of discipline, of
training, of obedience, and of learning the secret of willingness to
die for a principle' (quoted according to Searle 1976: 37).
This discussion among eugenicists was influenced by disagreement
about whether acquired traits could be inherited. The 'eugenic
militarists' often based their argument on the Lamarckian assumption
that the impact of the environment could improve the gene structure
of human beings. Their opponents focused their criticism on this
premise, which was becoming increasingly discredited among scientists.
The American eugenicist, Roswell H. Johnson, claimed that only 'by
a strange confusion of cause and effect' had it been assumed in some
quarters that the 'waste of virility from war' could be repaired by
universal military drill. He categorically denied that physical and
mental vigor increased by training would be passed on to future
offspring. His colleague Jordan stated that traits desirable in the
soldier such as physical strength, agility, courage and patriotism were
lost in the race which enforced the destruction of the soldierly: 'The
delusion that war in one generation sharpens the edge of warriorhood
in the next generation, has no biological foundation. It is the man
who is left who always determines the future' (Jordan 1910: 96).
However, despite the decline of Lamarckian thinking at the beginning
of the twentieth century, it was only because of the devastating
consequences of the First World War that eugenicists from different
countries developed a common position towards war.

The Impact of the First World War

Because of the controversy about the eugenic and dysgenic
effects of war, the eugenics societies in Germany, France, Great Britain
and the United States refrained from formulating a unified position
on this question. It was not until the outbreak of the First World War
that this situation changed. Although the different national societies

accepted their patriotic obligations, more and more of their members started to worry about the 'dysgenic' effects of the war. The *Eugenics Review* (7, 1915: 131) remarked that in Great Britain the subject of 'eugenics and the War' was treated 'in many parts of the country and by different speakers'. In its pages, Edward B. Poulton (1916: 39–40) and Leonard Darwin (1917), president of the British eugenics society, agreed that 'war unquestionably killed off the better types, and was therefore highly dysgenic'. Poulton was especially concerned that 'the young men who have willingly gone forth from Oxford and from Cambridge for their country and for the liberty of the world' were dying in the trenches. Because their courage was, 'intellectual and moral rather than physical' these were the men needed for the coming 'social reconstruction' (see also: *Eugenics News*, 1, 1916: 43–4). In the United States, the outbreak of the First World War motivated many eugenicists to link themselves with efforts for a fast ending of the war. Johnson (1915: 548) summarized the position of many of his colleagues in the American *Journal of Heredity*. He stated that because of the war the 'inherent quality' of the human species declined 'faster than in any previous similar length of time'. When, in 1917, Roland Campbell Macfie claimed that the killing of 500 of the best individuals would lead to an improvement of the stock because women would have greater choice among men he was immediately criticized by several of his colleagues.[3] In Italy and Austria eugenicists also became critical of the war. The Italian Marcello Boldrini saw three reasons for the 'racially damaging' effects of the war. First of all, the people fighting in the trenches were lost to the selection process. Secondly, the people of low physical and mental calibre who had been rejected for military service became fathers. Thirdly, because of the war, tuberculosis, malaria and mental illness could spread (Boldrini 1921; also Sergi 1917; *Archiv für Rassen- und Gesellschaftsbiologie* 14, 1921: 228; *Eugenics Review* 10, 1918: 113). In Austria, the anatomist Julius Tandler described the war as a 'monumental concentration of the struggle for existence'. He stressed the negative effects of the 'widespread mixture of races' which was an indirect consequence of the war. (Tandler 1916; see also Byer 1988: 73–5) In Germany, race hygienists had been strongly influenced by imperialist and militarist thinking. However, in the course of the war, their position changed fundamentally. Ernst Haeckel, honorary member of the German Society for Race Hygiene and before 1914 a glorifier of selection through war, was shocked by modern warfare:

The longer the terrible war of the nations lasts and the greater the values which it destroys in human lives, in cultural acquisitions and in material possessions, the more urgent grows the desire on all sides for the immediate establishment of peace. .... Our aim is to prevent the inevitable ... 'competitive struggle' from degenerating into a bloody and murderous 'struggle for existence'. The higher civilized nations should exercise mutual tolerance towards each other and combine for higher common cultural work in the service of true humanity. (Haeckel 1916: 104–5; see also Semmel 1958: 123)

Géza von Hoffmann (1916), the Hungarian link between the German race hygienists and American eugenicists, regretted that a considerable part of 'the best, the most courageous and the healthiest had been eradicated forever'. At the end of the war, the *Archiv für Rassen- und Gesellschaftsbiologie* assessed this change in thinking among German race hygienists. It claimed that even the race hygienists who had stressed the positive effects of wars before 1914 no longer denied the devastating contra-selective results of modern warfare any longer (Schweisheimer 1918/21: 11; see also Propping/Heuer 1991).

Ironically, the First World War brought eugenicists from different countries closer together. They intensified their informal contacts in spite of the fact that international gatherings ceased to take place between 1914 and 1918 and that the Permanent International Eugenics Committee stopped its activities during this time (Laughlin 1934: 2). They corresponded, reviewed each other's work and discussed their concerns about the dysgenic effects of the war. Eugenicists from different countries started to meet again immediately after the end of the war. Only the German and Austrian race hygienists and the Russian eugenicists were excluded from this. The first informal meeting took place in January 1919 and the first meeting of the Permanent International Eugenics Committee took place in October 1919. Eugenicists from the United States, Belgium, Great Britain, Australia, Denmark, France, Italy and Norway agreed to hold an international congress of eugenics as soon as possible. When in September 1921 more than 300 participants gathered in New York for the Second International Congress for Eugenics, the dysgenic effects of the World War was one of the major topics. The invitation for the Congress stated:

Since the First International Congress the world war has come and gone and the question in more than one country is whether the finest racial stocks have not been so depleted by it that they are in danger of extinction ... The war left the economic, sociological, and biological conditions of the world greatly disturbed. Never before has the need of international cooperation and enlightenment been felt so keenly. (*Eugenic News* 5, 1920: 12)

Eugenicists did not hesitate to claim that they had a direct contribution to make towards 'securing the peace which the leading civilized nations are anxious to obtain for the world' (Bedwell 1923: 429). At his opening address, the American eugenicist Henry Fairfield Osborn (1921: 311), president of the congress, claimed that there had never before been a moment in the world's history when an 'international conference on race character and betterment' had been more important. Europe, he continued, had lost, in 'patriotic self-sacrifice', the heritage of centuries of civilization which can never be regained. In certain parts of Europe 'the worst elements of society' had gained the ascendancy and threatened the 'destruction of the best'.

Despite the desperation about the devastating effects of the war, eugenicists saw an extraordinary chance for developing and propagating their eugenic programme. Paul Popenoe, an influential eugenicist from California, stated that the war had forced people to think about 'race value' and 'artificial selection' and that eugenic thinking had gained a new popularity in the United States (Popenoe 1923/1924: 196; see also Popenoe/Johnson 1933: v). In Germany, as the historian Paul Weindling has pointed out, virtually every aspect of eugenic thought and practice – from 'euthanasia' of the unfit and sterilization to positive welfare-developed between 1918 and 1924 (Weindling 1989: 307).[4]

The Eugenically Minded Peace Order and the Question of the 'Inferior Members' of Society

Impressed by the devastating effects of the First World War, eugenicists agreed on two ways of addressing the dysgenic effects of war. First, eugenicists saw themselves as obliged to try and prevent another war. Ignaz Kaup, an influential race hygienist from Munich, claimed the fashionable flirtation with the idea of 'the wild struggle for existence' represented a serious danger for civilization. He demanded that German race hygiene should break with this thinking 'once and for all'. (Kaup 1922: 15) The programme commission of the Eugenics Society of the United States claimed that the effort to prevent future wars was 'a matter of fundamental eugenic concern' (see *Eugenic News*, 8 (1923): 72). Secondly, eugenicists understood the need to make up for the loss of 'valuable stock' during the war. In 1917, the Berlin Society for Race Hygiene had already presented a memorandum to the Reichstag in favour of making medical examination compulsory before marriage. With this 'certification of fitness' German race hygienists wanted to reduce the dysgenic effects of the war (Ellis 1919: 110). Georges Papillault (1921; see also *Eugenic News*, 7, 1922: 6) claimed before the French Society for Eugenics that the Great War had confirmed the laws of eugenics and that the reproduction of 'inadequates' had to be prevented more urgently than ever before.

Eugenicists linked their peace policy directly with the question of the so-called 'inferior members' of society. Especially after the First World War they saw warfare and welfare as directly intertwined. Irving Fisher, a prime mover in the American eugenics movement, demanded, as immediate consequences of the war: first a league of nations to prevent another war, and secondly the prevention of inmates of mental hospitals from procreating. The British eugenicist Havelock Ellis (1919: 120–1) claimed that the war has rendered the 'relation of the fit members of the community to the unfit' far more acute: 'Never before has it been so urgent a demand on us to do all in our power to prevent the breeding of the unfit and to limit the breeding of the less fit members of society.'

The strategy which eugenicists used to link their peace policy with the question of the unfit was the transfer of the selection process from the level of the group or state to that of the reproductive cells (Weindling 1989: 125). They stressed that the struggle for existence did not cease to exist with a eugenically minded peace policy. It would only be more rationally planned. The process of selection and survival of the fittest would be transferred to the level of the individual. In the thinking of eugenicists, the systematically organized struggle for existence was much more promising than the wild fights favoured by social Darwinists in the second half of the nineteenth century. Fritz Lenz stressed that collective or group selection brought positive results only among primitive people. With the modernization of society, group selection through wars would lose its positive effects. Therefore, it was fundamental to move from that level to a systematic selection on the individual level (Lenz 1923: 53; see also Ploetz 1895: 230).

Not without reason were the mentally handicapped and the 'fight' against them now described with military analogies. After the First

World War, the eugenicists saw their countries invaded by an 'army of the unfit'. The social democratic eugenicist Alfred Grotjahn from Berlin saw the isolation of 'the army of beggars, alcoholics, criminals, prostitutes, psychopaths, epileptics, mental invalids, feebleminded, and cripples' as central for the recreation of the German people. The American eugenicist Edward Grant Conklin claimed that the 'armies of defective and delinquent persons in every nation and race' testified to the fact that 'there is an urgent need for racial improvement'.[5] Herman Lundborg, the leading race hygienist from Sweden, described the 'internal enemies' – the 'inferior' members of society – as extremely dangerous because they would be responsible for the degeneration of a race. He demanded that doctors and sociologists should lead the fight against them (Lundborg 1926: 3).

The First World War influenced the discussion about the medical killing of 'lives not worth living'. The debate about the so-called 'euthanasia' of mentally handicapped people which began in Germany immediately after the war was dominated by the contrasting images of 'valuable' soldiers dying in the trenches and 'lives not worth living' vegetating in mental institutions (Schmuhl 1987: 107). The German lawyer Karl Binding, one of the main proponents of the killing of the mentally handicapped, wrote that he was deeply disturbed by the 'sharp discord' between a 'battlefield full of thousands of dead youths' and 'mental institutions with their care for their living inmates'. The psychiatrist Alfred Hoche with whom in 1920 Binding published an influential book about the 'destruction of life not worth living' changed his opinion regarding 'euthanasia' only after the experience of the First World War. Germany's defeat in the war and the loss of one of his own sons made him into one of the most aggressive promoters of the killing of mentally handicapped people (Binding/Hoche 1920; see also Lifton 1986: 47; Weindling 1989: 394–5).

Interestingly enough, the debate about 'euthanasia' took place mostly outside the German race hygiene movement (see Weingart/Kroll/Bayertz 1988: 524). Also the eugenics movements in Great Britain, France and the United States did not become active in the discussion of the killing of 'lives not worth living'. Fritz Lenz (1932: 307) wrote in the main race hygienist textbook that 'euthanasia' was from a eugenic point of view not very effective and that, therefore, race hygienists should not support corresponding initiatives. The American *Eugenic News* claimed that there were only small practical applications of eugenics in euthanasia. It warned against placing them together in the same programme of social reform: 'For the ancient

Spartan and for the animal-breeding world euthanasia is a practical technique for breed-improvement, but in eugenics mankind has something more basic, less cruel and much more effective for purging racial and family stocks of degenerate qualities' (*Eugenic News* 20 (1935): 38–9).

However, the race hygienists and eugenicists agreed with some of the basic assumptions of the proponents of euthanasia. They both condemned the disastrous effects of the World War and they both tried to prevent so-called 'inferior life'. Therefore, eugenicists generally did not attack Binding, Hoche and their supporters on the grounds that they were supporting the murder of human beings, but instead argued that a 'selection process' should terminate the production of 'inferior' offspring and not the existence of already living people. The German eugenicist Karl H. Bauer (1926: 27) stated that in the selection process the death of the individual is not of central importance, because 'we all must die'. Rather the 'number and hereditary value of the offspring' should be the central focus.

The International Organization of Eugenicists against War

In the 1920s the informal international contacts among eugenicists dealing with the eugenic and dysgenic consequences of war became more and more institutionalized. In 1927 the International Federation of Eugenic Organizations – the successor of the Permanent International Eugenics Committee – decided to form an international committee on eugenics and war. The initiative came from American eugenicists, particularly from Charles Davenport and Irving Fisher. Davenport, who was president of the International Federation, thought that a strong committee on this subject might influence governments in their attitude towards warfare.[6] The Italian eugenicist, Corrado Gini, who had strong links to a special bureau at the Italian Ministry of War, became chairman of the committee and was responsible for coordinating research on the eugenic and dysgenic effects of the First World War. The committee, which consisted of eugenicists from France, Germany, the Netherlands, Italy, Belgium, Japan, Great Britain, Bulgaria, India, Hungary, Austria and the United States, planned a systematic investigation in every nation that had participated in the First World War.[7] Although the committee never succeeded in presenting a common report, its members presented their investigations at several international gatherings. The German population scientist, Friedrich Burgdörfer, and the French eugenicist, Henry Briand, gave talks at the International Congress of Population Science in Rome 1931. At the 1932 International Congress for Eugenics in New York, Harrison Hunt (1934: 244; see also Hunt 1930), one of the American committee members, claimed that his research had illustrated the dysgenic effects of war. Theodore Szél (1934: 252), a eugenicist from Hungary, used the same occasion to state that there could be no doubt that the 'eugenic effects of the World War, which in certain respects were beneficial, had become completely dwarfed by the dysgenic effects'. Gini (1934: 239) himself gave a differentiated picture of the eugenic and dysgenic effects. He concluded, however, that the 'selection which occurred among the soldiers during the war period has had an unfavourable effect from the eugenic point of view'.

With the growing international tensions of the 1930s the international eugenics movement became more and more active in propagating a eugenic peace order. In 1934 eugenicists from twelve countries met in Zurich to discuss recent developments in eugenics. Ernst Rüdin, successor of Davenport as president of the International Federation of Eugenic Organizations, welcomed the participants and stressed that the 'will for peace between the people' was an important 'common tie' between eugenicists from all nations. He stated that 'all eugenicists know that war would mean an awful eradication of the most capable and valuable elements of a nation' (International Federation of Eugenic Organizations 1934: 4). Besides an evaluation of the new German race policies, the potentially negative effects of a new world war were the main topic at the conference. Initiated by Alfred Ploetz, the conference passed a resolution expressing its participants' worries about the menace of a new war. The resolution claimed that a new war would again kill the most capable men and that this loss of capable 'human material' could be disastrous for the western world (see *Journal of Heredity*, 26, 1935: 10). The resolution was sent to the prime ministers of all the principal governments.[8]

The Radicalization of the Eugenically Motivated Peace Policy during National Socialism and the Killing of the Mentally Handicapped

The declaration of the conference in Zurich was an important victory for the participating German race hygienists. They succeeded in linking a positive reaction to Nazi race policies to a resolution condemning wars as highly contra-selective. Several Nazi journals and Nazi officials therefore welcomed the results of the conference (see *Volk und Rasse*, 8, 1934: 298; *Rassenpolitische Auslands-korrespondenz*, 4/1934). When in 1935 an international congress for population science took place in Berlin, the Nazi government tried to take up the peace resolution made in Zurich one year before. Ploetz discussed a possible 'anti-war resolution' with the Minister of the Interior, Wilhelm Frick.[9] Eventually the congress did not pass a resolution similar to the one in Zurich, but Ploetz gave a talk in which he condemned war as one of the most disastrous environments for race hygiene (Ploetz 1936: 619). The Nazi government hoped to reward Ploetz's efforts for a 'eugenic peace order' by supporting his candidature for the Nobel peace prize. Several Scandinavian eugen-icists, like Alfred Mjöen and Hermann Lundborg, nominated him unsuccessfully for the prize in 1936.[10]

The Nazi race politicians completely adopted the eugenic argument against war. Their propaganda claimed that German race policies were central for ensuring peace among the different nations.[11] Walter Gross, head of the Racial Political Office of the NSDAP, speaking before diplomats, described 'race policy as peace policy'. The aim to improve the German race would force Nazi Germany to be a peaceful nation: 'Because Nazi Germany thinks racially, it wants peace. The National Socialist ideology represents the most peaceful one, because it is the only one which sees its aim as the preservation of the racial essence of the people.' He concluded that even a victory in a potential war would be a defeat biologically (Gross 1935: 1–6). The race hygienists in Nazi Germany developed the vision that only a com-munity of healthy people could develop a stable peace order. In the second edition of the commentary to the German sterilization law, Arthur Gütt, Ernst Rüdin and Falk Ruttke expressed their hopes that 'Germany's struggle for hereditarily healthy offspring' would lead to 'true community of the healthy and strong people'. Only this community would be able to 'give the world a new and better form' and would result in the ‚true peace‘ among the most capable (Gütt/ Rüdin/Ruttke 1936: 72; see also Steinwallner 1937: 251). Hitler himself stood for this vision of 'peace among the selected people'. He took over the eugenic claim that every war would only destroy the most valuable and that therefore the National Socialists' will for peace was their 'deepest ideological conviction' (see Frercks 1937: 45–6).

This propaganda for 'race policy as peace policy' was the ideological matrix within which the Nazis could justify the killing of the handicapped. The specific connection made between the extraordinary situation of war and the 'special' sacrifice of the mentally handicapped was already obvious in the discussion about compulsory sterilization. Gross (1935) justified mass sterilization in Nazi Germany by the fact that a state demanding the lives of its soldiers could also demand from certain people that they should give up their right to procreate. Ruttke (1934) claimed that the Germans had seen more than once that for the 'public welfare' the state had asked its 'best citizens' to sacrifice their lives. Therefore, it would be 'strange' if it could not ask a much smaller sacrifice from 'hereditarily inferior people'. When in 1935 Eugen Stähle (1935: 1; see Bock 1997), an administrator in the Ministry of the Interior of Württemberg, had to explain the deaths resulting from compulsory sterilization, he defended them by comparing them to the soldiers dying in the First World War. Later, Stähle who was active in carrying out the murder of mentally handicapped people, used the same argument to justify the Nazi killing programme:[12] 'If during the war we ask thousands of young and healthy people to sacrifice their lives for the community, we can ask the same sacrifice from the incurably ill.' After 1945, Hermann Pfannmüller, another central figure in the killing programme, justified his participation in a very similar way. He put on record that he just could not bear the fact that 'the best, the flower of our youth' lost their lives at the front 'in order that feebleminded and irresponsible asocial elements could have a secure existence in the asylums' (quoted according to Schmidt 1965: 34). This 'deeper psychological relationship between "euthanasia" and war' had a strong influence on the killing process (Lifton 1986: 63). Already in 1935 Hitler was said to have stated that war was the best opportunity for the 'elimination of the incurably ill'.[13]

The outbreak of the Second World War gave Hitler immediate cause for launching a programme for the systematic extermination of the mentally handicapped. In his notorious 'euthanasia' decree he empowered Philipp Bouhler and Karl Brandt to administer the killing programme (translation according to Lifton 1986: 63): 'Reich Leader Bouhler and Dr. Brandt are charged with the responsibility for expanding the authority of physicians, to be designated by name, to the end that patients considered incurable according to the best available human judgement of their state of health, can be granted a mercy death.'

Although the decree was actually issued at the end of October 1939 it was symbolically back-dated to 1 September, the day when Nazi Germany invaded Poland. The association of the killing programme with the outbreak of the Second World War was more than clear. Hitler and his colleagues took up the connection between war and peace and the question of the 'inferior members' of the German people.

German race hygienists were confronted with the dissociation of their peace and race policy. For them the apparent rationale, system and logic of race policies during peacetime seemed to end. They felt that the Second World War interrupted their biological mission and that, therefore, extraordinary measures were justified. Already in 1935 Alfred Ploetz had claimed that in the case of war the state had to make up for war's contra-selective effects through the 'increase of the extermination and selection quotas' (Ploetz 1936: 618). Rüdin stated at the end of 1939 that the English government had begun the war despite the efforts of German race hygienists and their colleagues in other countries to prevent a war between the European nations. So Germany had to fight back and at the same moment continue its race hygienist mission (Rüdin 1939: 443–5; see also Weber 1993: 235). Hermann Ernst Grobig, one of Rüdin's collaborators at the German Institute for Psychiatry in Munich, claimed in 1943 that the 'race hygienist and race political measures should in no way take a second place behind the war efforts'. On the contrary, because of the war, the measures for race improvement had to be intensified. This strategy, Grobig stated, was central not so much for the outcome of the war as for the 'consolidation of the victory' (Grobig 1943; see also Weber 1993: 268).

Similar arguments were made by the population scientist Friedrich Burgdörfer. In 1942, he argued that from a eugenic point of view the war was disastrous not only for the German people but also for the English and French 'who had already been biologically on a steeply sloping road' (Burgdörfer 1942: 5). He described it as a patriotic obligation for the Germans to improve their racial values. Every 'hereditarily healthy' couple that neglected its reproductive obligations should be held responsible for 'national desertion'. Burgdörfer added that, besides winning the war, the preservation and increase of 'people power' should be a principal goal. Only this way could Germany pass the 'biological endurance test' (Burgdörfer 1942: 29, 39).

It is in this context that we must view the attitudes of influential eugenicists like Otmar von Verschuer, Eugen Fischer, Fritz Lenz and

Ernst Rüdin towards the mass killing. The contrast between their utopian vision of a eugenic peace order of superior human beings and the devastating results of the Second World War made them completely immune to moral scruples about the killing of mentally handicapped people. They saw the necessity not only for an economic and military mobilization, but especially for a biological one. This could only mean a further radicalization of their race hygiene policy. The killing programme was the symbiosis of an economic, military and race hygienic mobilization at the 'home front'. In the perverse logic of race hygienists the killing programme helped to save economic resources, created hospital beds for injured soldiers and could counteract the supposedly 'racial degeneration' of the German people.

Conclusion

The leading eugenicists in Nazi Germany did not object to killing in principle. The starting point for their peace policy was not a rejection of killing for humanitarian reasons but the contra-selective effects of certain forms of killings. For them killing was a neutral issue subordinated to the higher goal of race improvement. Therefore, eugenicists could imagine a eugenically perfect war. They could consider wars between primitive people as positive from a eugenic point of view. They thereby distinguished themselves from all other pacifist movements in the early twentieth century. For eugenicists, the outbreak of the Second World War meant the destruction of a proper eugenic situation. They saw their utopian ideals, which seemed to come true under the Nazis, destroyed by the war. In this situation they considered extraordinary means to be legitimate. In this light the death of tens of thousands of mentally handicapped people was partly due to the unfulfilled utopian vision of a 'eugenic peace' among peoples of superior racial stock.

Notes

1. For a good introduction to the problems of the historiography of the killing of 'lives not worth living', see Nowak (1988) and Burleigh (1991 and 1994).

2. I am dealing with the position of eugenicists of different nationalities. I am aware that this transnational approach is dangerous because it risks linking developments in the different national movements too closely. I try to avoid this danger by focusing primarily on the interaction of eugenicists from different national backgrounds and on the common positions within the international eugenic movement. In using this discourse analysis method, I reconstruct the inner logic of the argumentations of eugenicists and link the arguments with their actual behaviour during the euthanasia action. For further details, see Bock 1986; Bock 1991; Kühl 1994.

3. See Macfie (1917a and b); see letter to the editor of Clifford Musprat, *The New Statesman*, 17 February 1917 and *Eugenic News*, 2, 1917: 89.

4. In a slightly different interpretation, I put this development down to the experience of the First World War and not, as does Weindling, to the difficulties of material day-to-day existence and the popular sense of a struggle for survival.

5. Edward Grant Conklin, *The Purposive Improvement of the Race*, Conklin Papers, Princeton University, Box 16, p. 1.

6. Davenport Papers, American Philosophical Society, letter from Davenport to Hodson, 23 February 1927.

7. Davenport Papers, American Philosophical Society, letter from Gini to Jungblut, 29 May 1927. The list of members is printed in *Eugenic News*, 15 (1930): 26.

8. Davenport Papers, American Philosophical Society, Hodson, memo dated 1934.

9. Ploetz Papers, Herrsching, Ploetz diary, 16 July 1935.

10. Ploetz Papers, Herrsching, Ploetz diary, 9 June 1936; see also Doelecke 1975: 109.

11. '"Rassismus" über Europa?', *Nationalsozialistische Parteikorrespondenz*, 26 August 1937.

12. Minutes of the examination from 26 June 1945; KS 6/49 StA Tübingen gegen Angeschuldigte im Grafeneck-Komplex. Quoted by Klee (1983: 90).

13. Testimony of Professor Böhm 12 July 1961; KS 2/63 GStA Frankfurt gegen Prof. Werner Heyde u.a., pp. 42–3. See Lifton (1986: 50).

References

Primary Sources

Conklin Papers, Princeton University, Princeton.
Davenport Papers, American Philosophical Society, Philadelphia.
Ploetz Papers, Private Archive of the Ploetz Family, Herrsching/ Ammersee.

Secondary Literature

Ammon, Otto (1895), *Die Gesellschaftsordnung und ihre natürlichen Grundlagen*, Jena: Fischer.

Bauer, Karl Heinz (1926), *Rassenhygiene. Ihre biologische Grundlagen*, Leipzig: Quelle & Meyer.

Bedwell, C. E. A. (1923), 'Eugenics in International Affairs', in *Eugenics in Race and State. Second International Congress of Eugenics, 1921*, vol. 2, Baltimore, 429.

Binding, Karl and Hoche, Alfred (1920), *Die Freigabe der Vernichtung lebensunwerten Lebens. Ihr Maß und ihre Form*, Leipzig: Meiner.

Bock, Gisela (1986), *Zwangssterilisation und Nationalsozialismus. Studien zur Rassenpolitik und Frauenpolitik*, Opladen: Westdeutscher Verlag.

—— (1997), 'Sterilization and "Medical" Massacres in National Socialist Germany: Ethics, Politics, and the Law', in Manfred Berg and Geoffrey Cocks (eds), *Medicine and Modernity. Public Health and Medical Care in Nineteenth- and Twentieth-Century Germany* (Publications of the German Historical Institute, Washington D.C.), Cambridge: Cambridge University Press, 149–72.

Boldrini, Marcello (1921), 'Some Dysgenical Effects of the War in Italy', *Social Hygiene*, 7: 265–78.

Burgdörfer, Friedrich (1942), *Kinder des Vertrauens. Bevölkerungspolitische Erfolge und Aufgaben im Großdeutschen Reich*, Berlin: Eher.

Burleigh, Michael (1991), 'Surveys of Development in the Social History of Medicine: III. "Euthanasia" in the Third Reich: Some Recent Literature', *Social History of Medicine*, 4: 317–28.

—— (1994), *Death and Deliverance. 'Euthanasia' in Germany c. 1900–1945*, Cambridge: Cambridge University Press.

Byer, Doris (1988), *Rassenhygiene und Wohlfahrtspflege. Zur Entstehung eines sozialdemokratischen Machtdispositivs in Österreich bis 1934*, Frankfurt a. M; New York: Campus.

Darwin, Leonard (1917), 'The Disabled Sailor and Soldier and the Future of Our Race', *Eugenics Review*, 9: 1–10.

Doelecke, Werner (1975), *Alfred Ploetz (1860–1940), Sozialdarwinist und Gesellschaftsbiologe*, Frankfurt a. M: unpublished dissertation.

Ellis, Havelock (1919), *The Philosophy of Conflict and Other Essays in Wartime*, London.

Eugenics Education Society (n.d.), *The First International Eugenic Congress*, London.

Frercks, Rudolf (1937), *Deutsche Rassenpolitik*, Leipzig: Reclam

Friedlander, Henry (1995), *The Origins of Nazi Genocide. From Euthanasia to the Final Solution*, Chapel Hill: University of North Carolina Press.

Gini, Corrado (1934), 'Report of the Committee for the Study of the Eugenic and Dysgenic Effects of War', in *A Decade of Progress in Eugenics. Scientific Papers of the Third International Congress of Eugenics held at the American Museum of Natural History, New York, August 21–23, 1932*, Baltimore: Williams & Willkins, 232–43.

Grobig, Hermann Ernst (1943), 'Warum Rassenhygiene im Krieg', *Archiv für Rassen- und Gesellschaftsbiologie*, 37: 175–9.

Gross, Walter (1935), 'Die Bevölkerungspolitik und Rassenpolitik des neuen Deutschlands', *Rassenpolitische Auslandskorrespondenz*, 3/1935: 1–6.

Gütt, Arthur, Rüdin, Ernst and Ruttke, Falk (1936), *Gesetz zur Verhütung erbkranken Nachwuchses vom 14. Juli 1934 nebst Ausführungsverordnungen*, 2nd edn, München: Lehmann.

Haeckel, Ernst (1916), *Eternity. World-War Thoughts on Life and Death, Religion, and the Theory of Evolution*, New York.

Haller, Mark H. (1963), *Eugenics. Hereditarian Attitudes in American Thought*, New Brunswick: Rutgers University Press.

Hoffmann, Géza von (1916), *Krieg und Rassenhygiene. Die bevölkerungspolitische Aufgabe nach München*, München: Lehmann.

Hunt, Harrison Randal (1930), *Some Biological Aspects of War*, New York.

—— (1934), 'Is War Dysgenic?', in *A Decade of Progress in Eugenics. Scientific Papers of the Third International Congress of Eugenics held at the American Museum of Natural History, New York, August 21–23, 1932*, Baltimore: Williams & Willkins, 244–8.

Inge, William R. (1913/14), 'Depopulation', *Eugenics Review*, 6: 261.

International Federation of Eugenic Organizations (1934), *Bericht über die 11. Versammlung der Internationalen Federation eugenischer Organisationen. Konferenzsitzungen vom 18. bis 21. Juli 1934 im Waldhaus Dolden, Zürich*, Zurich.

Johnson, Roswell H. (1915), 'Natural Selection in War', *Journal of Heredity*, 6: 546–8.

Jordan, David Starr (1910), 'War and Manhood', *Eugenics Review*, 2: 95–109.

—— (1913), 'The Eugenics of War', *Journal of Heredity*, 4: 140–7.

—— (1915), *War and the Breed*, Boston.

Kaup, Ignaz (1922), *Volkshygiene oder selektive Rassenhygiene*, Leipzig.

Kellogg, Vernon L. (1912), 'Eugenics and Militarism', in Eugenics Education Society (ed.), *Problems in Eugenics. Papers Communicated to the First International Eugenics Congress 1912*, London, 220–31.

—— (1913), 'Eugenics and Militarism', *Atlantic Monthly*, 112: 99–108.

—— (1914), 'The Bionomics of War: Race Modification by Military Selection', *Social Hygiene*, 1: 44–52.

Klee, Ernst (1983), *'Euthanasie' im NS-Staat. Die 'Vernichtung lebensun-werten Lebens'*, Frankfurt a. M.: Fischer.

Kröner, Hans-Peter (1980), *Die Eugenik in Deutschland von 1891–1934*, Münster: Ph.D. at the University of Münster

Kühl, Stefan (1994), *The Nazi Connection. Eugenics, American Racism and German National Socialism*, New York/Oxford: Oxford University Press.

—— (1997), *Die Internationale der Rassisten. Aufstieg und Niedergang der internationalen Bewegung für Eugenik und Rassenhygiene im 20. Jahrhundert*, Frankfurt a. M./New York: Campus.

Laughlin, Harry H. (1934), 'Historical Background of the Third International Congress of Eugenics', in *A Decade of Progress in Eugenics. Scientific Papers of the Third International Congress of Eugenics held at the American Museum of Natural History, New York, August 21–23, 1932*, Baltimore: Williams & Willkins, 1–14.

Lenz, Fritz (1914/15), 'Eugenic Peace', *Archiv für Rassen- und Gesellschaftsbiologie*, 11: 558–9.

—— (1923), *Menschliche Auslese und Rassenhygiene*, 2nd edn, München: Lehmann.

—— (1932), *Menschliche Auslese und Rassenhygiene (Eugenik)*, 4th edn, München: Lehmann.

Lifton, Robert Jay (1986), *The Nazi Doctors. Medical Killing and the Psychology of Genocide*, New York: Basic Books.

Lundborg, Herman (1926), 'Die drohende Entartung gewisser Kulturvölker', *Zeitschrift für Volksaufartung und Erbkunde*, 1: 3–7.

Lutzhöft, Hans-Jürgen (1971), *Der nordische Gedanke in Deutschland 1920 bis 1940* (Kieler historische Studien 14), Stuttgart: Klett.

Macfie, Roland Campbell (1917a), 'The Selective Effects of War', *The New Statesman*, 8: 441–2.

—— (1917b), 'Some of the Evolutionary Consequences of War', *Science Progress*, 132–7.

Massin, Benoit (1996), 'L'euthanasie psychiatrique sous le IIIe Reich. La question de l'eugénisme', *L'information psychiatrique*, 8/199: 811–22.

Melville, C.H. (1910/11), 'Eugenics and Military Service', *Eugenics Review*, 2: 53–60.

Mjöen, Jon Alfred (1914), *Racehygiene*. Kristiania.

Müller-Hill, Benno (1984), *Tödliche Wissenschaft. Die Aussonderung von Juden, Zigeunern und Geisteskranken, 1933–1945*, Reinbek bei Hamburg: Rowohlt.

Nowak, Kurt (1988), 'Sterilisation und "Euthanasie" im Dritten Reich. Tatsachen und Deutungen', *Geschichte in Wissenschaft und Unterricht*, 39: 327–41.

Osborn, Henry Fairfield (1921), 'The Second International Congress of Eugenics. Address of Welcome', *Science*, 54: 311–13.

Papillault, Georges (1921), 'Conséquences psycho-sociales de la dernière guerre au point de vue eugénique', *Eugénique*, 2: 251–73.

Pearson, Karl (1905), *National Life from the Standpoint of Science*, London.

Ploetz, Alfred (1895), *Die Tüchtigkeit unserer Rasse und der Schutz der Schwachen*, Berlin: Fischer.

—— (1936), 'Rassenhygiene und Krieg', in Hans Harmsen and Franz Lohse (eds), *Bevölkerungsfragen. Bericht des Internationalen Kongresses für Bevölkerungswissenschaft Berlin, 26. August–1. September 1935*, München: Lehmann, 615–20.

Popenoe, Paul (1923/1924), 'Rassenhygiene (Eugenik) in den Vereinigten Staaten', *Archiv für Rassen- und Gesellschaftsbiologie*, 15: 184–93.

—— and Johnson, Roswell H. (1933), *Applied Eugenics*, 2nd edn, New York: Macmillan.

Poulton, Edward B. (1916), 'Eugenics Problems after the Great War', *Eugenics Review*, 8: 34–49.

Propping, Peter and Heuer, Bernd (1991), 'Vergleich des "Archiv für Rassen- und Gesellschaftsbiologie" (1904–1933) und des "Journal of Heredity" (1910–1939)', *Medizinhistorisches Journal*, 25: 78–93.

Reyer, Jürgen (1991), *Alte Eugenik und Wohlfahrtspflege. Entwertung und Funktionalisierung der Fürsorge vom Ende des 19. Jahrhunderts bis zur Gegenwart*, Freiburg: Lambertus.

Rüdin, Ernst (1939), 'Der uns aufgezwungene Krieg und die Rassenhygiene', *Archiv für Rassen- und Gesellschaftsbiologie*, 33: 443–5.

Ruttke, Falk (1934), 'Erbpflege in der deutschen Gesetzgebung', *Ziel und Weg*, 4: 600–3.

Schmidt, Gerhard (1965), *Selektion in der Heilanstalt 1939–1945*, Stuttgart: Evangelisches Verlagswerk.

Schmuhl, Hans-Walter (1987), *Rassenhygiene, Nationalsozialismus, Euthanasie. Von der Verhütung zur Vernichtung 'lebensunwerten Lebens', 1890–1945* (Kritische Studien zur Geschichtswissenschaft 75), Göttingen: Vandenhoeck & Ruprecht.

—— (1997), 'Eugenik und "Euthanasie" – Zwei Paar Schuhe? Eine Antwort auf Michael Schwartz', *Westfälische Forschungen*, 47: 757–62.

Schuster, Edgar (1912), *Eugenics*, Oxford.

Schwartz, Michael (1996), '"Rassenhygiene, Nationalsozialismus, Euthanasie"? Kritische Anfragen an eine These Hans-Walter Schmuhls', *Westfälische Forschungen*, 46: 604–22.

—— (1998), '"Euthanasie"-Debatten in Deutschland (1895–1945)', *Vierteljahrshefte für Zeitgeschichte*, 46: 617–65.

Schweisheimer, W. (1918–21), 'Bevölkerungsbiologische Bilanz des Krieges 1914/18', *Archiv für Rassen- und Gesellschaftsbiologie*, 13: 11.

Searle, Geoffrey R. (1976), *Eugenics and Politics in Britain, 1900–1914*, Leiden: Noordhoff.

Semmel, Bernard (1958), 'Karl Pearson: Socialist and Darwinist', *British Journal of Sociology*, 9: 111–25.

Sergi, Giuseppe (1917), *La guerra e la preservazione della nostra stirpe*, Rome.

Smith, Frank O. (1914), *Eugenic Peace*, Washington.

Stähle, Egon (1935), No title, *Ärzteblatt für Württemberg und Baden*, 7: 1.

Steinwallner, Bruno (1937), 'Rassenhygienische Gesetzgebung und Maßnahmen ausmerzender Art', *Fortschritte der Erbpathologie, Rassenhygiene und ihrer Grenzgebiete*, 1: 193–260.

Szél, Theodore (1934), 'The Genetic Effects of the War in Hungary', in *A Decade of Progress in Eugenics. Scientific Papers of the Third International Congress of Eugenics held at the American Museum of Natural History, New York, August 21–23, 1932*, Baltimore: Williams & Willkins, 252.

Tandler, Julius (1916): 'Krieg und Bevölkerung', *Wiener klinische Wochenschrift*, 29: 499–504.

Thomson, J. Arthur (1915), 'Eugenics and War', *Eugenics Review*, 7: 1–14.

Weber, Matthias M. (1993), *Ernst Rüdin. Eine kritische Biographie*, Berlin: Springer.

Weindling, Paul (1989), *Health, Race and German Politics between National Unification and Nazism, 1870–1945*, Cambridge: Cambridge University Press.

Weingart, Peter, Kroll, Jürgen and Bayertz, Kurt (1988), *Rasse, Blut und Gene. Geschichte der Eugenik und Rassenhygiene in Deutschland*, Frankfurt a. M.: Suhrkamp.

Wiggam, Albert Edward (1923), *The New Decalogue of Science*, Indianapolis.

JOSEF REINDL

Believers in an Age of Heresy? Oskar Vogt, Nikolai Timoféeff-Ressovsky and Julius Hallervorden at the Kaiser Wilhelm Institute for Brain Research*

During the 1930s and until the opening of the National Institute of Mental Health in Bethesda, USA, soon after the Second World War, Berlin-Buch was the world's largest brain research institute (Richter 1996: 384). The remains of the institute were taken over by the Soviets and then given to the East German Academy of Sciences in 1947. From then until the collapse of the GDR, Berlin-Buch housed the largest biomedical research complex of the Academy and was one, if not the most important, centre of East German biomedical research. The work undertaken there focused on basic and clinical research into cancer and heart disease. After reunification a large-scale federal research institute was established in Berlin-Buch, the Max Delbrück Centre for Molecular Medicine (Bielka 1997: 48–94, 107–27).

This essay looks at three scientists working at the Kaiser Wilhelm Institute for Brain Research in Berlin during the Third Reich.* Focusing on their personal attitudes, research interests and methods, as well as their perception of science's role in society, I intend to address two main questions:

* This article draws on results of the author's forthcoming study *The History of Biomedical Research at Berlin-Buch. 1920's to Today*.

211

- First, what continuities and discontinuities can be recognized before and after 1933?
- Secondly, how should we judge the scientists' attitudes, work and activities during the Third Reich?

What are the advantages of a biographical approach when dealing with science? It allows us a detailed look at the micro-level of science, or in other words, at the most basic decision-making process in any kind of research: in which fields of inquiry does the scientist choose to engage, with which methods and intentions, and to what end? This is particularly interesting in the case of the three scientists presented here who worked in a period of far-reaching political change but at the same time regarded themselves as apolitical. None of them had any official functions in National Socialist science policy and only one of them joined the party – in 1937 – and even then remained politically inactive. This leads us to the question: did the three remain what could be called 'Believers in an Age of Heresy' in the sense that they focused exclusively on the 'pure' progress of science? Or did they allow their research to be used and abused for broader political purposes?

Oskar Vogt

Oskar Vogt was born in 1870 in Husum. He studied medicine and started to work as a neurologist and psychiatrist in Berlin in 1898. In the same year he and his wife Cécile founded a private institute to undertake research on brain anatomy and physiology. This institute was incorporated into the Kaiser Wilhelm Society in 1915 with the help of Gustav Krupp von Bohlen und Halbach, with whom Vogt, being the private psychiatrist of the Krupp family, had close personal connections. Vogt was director of this institute until 1937 when he was dismissed after serious conflicts with the National Socialists. It is generally agreed that this was a result of his anti-Nazi stance, his personal integrity and his humanitarian attitude (Haymaker 1951: 180–4, 195; Kirsche 1986: 5–11, 16–17, 20; Richter 1996: 352–9).

This judgement is based on the events which took place in spring and summer 1933 when his institute was raided twice by the SA. Several employees were taken into custody for a few days and severely maltreated. Vogt was accused of having defamed National Socialism, of employing Jews, and of tolerating 'communist propaganda' at his

Figure 10   Oskar Vogt (1870–1959).

institute. Subsequent Gestapo investigations ended with the result that the first two charges could not be upheld. But he was blamed for tolerating communist activities because members of this party had been employed at his institute.[1] As a consequence Vogt was replaced in 1937 by Hugo Spatz, hitherto at the German Research Institute for Psychiatry (Deutsche Forschungsanstalt für Psychiatrie) (Haymaker 1951: 195–6; Kirsche 1986: 17–18; Richter 1996: 388–92).

Max Planck, President of the Kaiser Wilhelm Society, and Vice-President Krupp von Bohlen und Halbach, tried in earnest to support

Vogt. Krupp even threatened to resign as head of the brain institute's advisory board, which he actually did after Vogt was dismissed.[2] Krupp subsequently financed a new brain research institute established in the Black Forest, which was led by Vogt until his death in 1959 (Haymaker 1951: 196–8; Kirsche 1986: 18–19; Richter 1996: 395–404).

A study of the archival sources on Oskar Vogt shows that he was, as suggested by most writers, an impressive although sometimes slightly difficult personality (Haymaker 1951: 198–200; Kirsche 1986: 16–17). Numerous incidents show that he provoked unnecessary and yet serious conflicts with colleagues, public administrators, and the leadership of the Kaiser Wilhelm Society. I use the following three examples to shed some light on his personality and the way in which he pursued his aims.

Since the incorporation of his institute into the Kaiser Wilhelm Society Vogt urged the construction of a new building and repeatedly threatened to resign as director if his demands were not met.[3] For years the Kaiser Wilhelm Society negotiated with various German and Prussian ministries to get the necessary funds, but this proved rather difficult due to the financial turmoil of the post-war years.[4] In the course of the 1920s the Kaiser Wilhelm Society was finally able to consolidate its financial base, which allowed the establishment of several new and the extension and modernization of existing institutes (vom Brocke 1990: 227–49, 292–314). In the late 1920s an amount of 3.5 million Reichsmark had finally been raised for Vogt's institute. It included 1.3 million Reichsmark from the Rockefeller Foundation, which during this decade heavily subsidized German biomedical research.[5] It also financed in these years new buildings for the German Research Institute for Psychiatry (Deutsche Forschungsanstalt für Psychiatrie) in Munich and a new neurologic research institute in Breslau, headed by the renowned Otfried Foerster, with whom Vogt cooperated closely (Holdorff 1989: 132; Richter 1996: 382–3; Weber 1991: 157; Weindling 1988: 128–35).

Although construction work had already begun in Buch on the outskirts of Berlin, the World Economic Crisis forced the Kaiser Wilhelm Society to curtail Vogt's annual budget, as well as that of all other institutes. As he had done on various occasions before, Vogt again threatened to resign, which at this time posed a threat to the Kaiser Wilhelm Society as a whole. As nobody could replace him, Vogt was urged to maintain his position because his withdrawal would have meant the end of the institute. Closing down an institute for which the Kaiser Wilhelm Society had just received large sums would have put the Society in an awkward position when applying for funds in the future. But Vogt still insisted on the increase of his budget to the pre-crisis level, and only Krupp's personal and repeated intervention forced him to scale down these demands. 6 On 2 June 1931, the newly constructed buildings were finally inaugurated in Berlin-Buch, giving a home to 100 employees including thirty scientists (Richter 1996: 382–4).

After the aforementioned SA raids Vogt pinned three announce-ments to the institute's notice board stating that anyone who questioned the director's competence would be dismissed immediately. This included informing non-employees about developments at the insti-tute, which had triggered the SA attacks. Vogt entitled these notes 'Information on behest of the General Administration of the Kaiser Wilhelm Society', which in fact was not true. On the contrary, only after he had already put up these announcements did he inform the Society's leadership of what he had done. The latter told him that his formulations were too drastic and that he should have informed the administration in advance.[7]

While trying to secure his position by referring to the Kaiser Wilhelm Society, Vogt attempted to convince the authorities that his research was in their interest. Presenting his annual report to the institute's advisory board on 17 July 1934 he emphasized: 'the institute did not have to change its research programme during the last year because its work has been in line with the demands of the new government ever since!' Vogt added that when studying the NSDAP's publications he recognized that 'the scientific aims of the party and the practical conclusions drawn from these' were identical with those of his institute. He also reported that due to a growing interest in 'questions of inheritance, the fight against degeneration and the struggle for an improvement of the species (*Aufartung*)' he had organized numerous public tours and talks at the institute. Vogt mentioned in particular visiting groups of the 'SA Reich Leader School' (Reichsführerschule der SA), whom he had tried to convince to support his case, which at this time was investigated by Gestapo.[8]

One can assume in Vogt's favour that statements like these were intended to calm down his opponents. Texts written by Vogt during the 1920s and early 1930s, newspaper articles reporting on his research, and other sources show clearly that he certainly had no political affiliations with National Socialism. But these texts also indicate that the aim and purpose of his brain research displayed an affinity with certain aspects of National Socialist policy. To elaborate on this, one has to turn to brain research in general and to the ideas pursued by Vogt in particular.

Oskar Vogt and his wife Cécile attempted to explore the physiological foundations of normal and pathological reactions of the brain. They concentrated on the brain's architecture, or in other words, on the composition, size, form and function of nerve cells in different fields of the cerebral cortex. Thereby Vogt and others tried to discover how various functions of the body were connected to specific brain areas. His aim was, in the words of Vogt, 'to find the brain-physiological counterpart for each mental and psychological element and condition' (Vogt, Oskar and Cécile 1919a: 285). This was based on his assumption that behaviour, talents and psychological diseases stemmed from particular characteristics of the brain's anatomy (Hassler 1959/60: 242–4, 247–9; Kirsche 1986: 8, 26–37).

To find the anatomical and physiological causes for talents and diseases, as well as for the individual's propensity to commit crime, Vogt and others examined the brains of scientists, artists, politicians and criminals (Finger 1994: 34–5, 306–8). In 1930, Vogt described the purpose of this research as follows: 'It serves to create an objective foundation for an inquiry into individuals "born criminal" (*konstitutionelle Verbrecher*). A precise knowledge of inferior and superior elements in the brain anatomy is thereby intended to create the basis for practical measures to suppress the inferior (*Unterdrückung des Unterwertigen*) and to improve the superior (*Höherzüchtung des Vollwertigen*).'[9]

The idea that the human brain could be improved rested on Darwin's concept of evolution, which regarded the development of species as the result of a continuous process of natural selection. Darwin's cousin Francis Galton applied this mechanism to the human race, a development which is generally described as social Darwinism. Galton argued that in civilized conditions mankind did not undergo natural selection and this led to the biological deterioration of the race. Therefore only birth and marriage control which allowed 'valuable individuals' to reproduce would ensure the improvement of humanity. He called this eugenics (Magner 1979: 393–5; Mayr 1984: 501–2; Porter 1990: 1039).

The zoologist Ernst Haeckel from Jena became the leading promoter of Darwinism, as well as social Darwinism, in Germany. Haeckel strongly emphasized the selection mechanism and attempted to construct an evolutionary theory on this basis (Magner 1979: 179–80; Weiss 1990: 13–14: Zmarzlik 1963: 249). As an adolescent Vogt had become interested in questions of evolution, and in 1890 went to Jena to study with Haeckel. Vogt had therefore already become familiar with the ideas of Darwinism and social Darwinism as a student (Haymaker 1951: 181; Kirsche 1986: 7).

Eugenics, for which in Germany the term 'race hygiene' (*Rassenhygiene*) was used, benefited from the so-called 'degeneration debate' which had started in the mid-nineteenth century. In this debate scientists argued that the large number of children in lower-class families and smaller numbers in the upper class led to a process of degeneration of industrial societies. This was based on the assumption that any progress made was dependent on the higher echelons of society. Using this logic it could be argued that 'degeneration' would prevent the advancement of societies. Taking place as it did in a period of increasing rivalries between nation-states it was not surprising that this mainly biomedical dispute gained political importance (Harrington 1987: 101–4; Pick 1989: 155–221; Weingart 1987: 159). It caused increasing public interest in questions of heredity, drawing on the increased interest in genetics which had started with the rediscovery of Mendel's Laws in 1900. At the same time genetics benefited from the high expectations society attached to its findings. The fear of being left behind prompted many countries to allocate substantial financial means to this field. This was demanded particularly by the promoters of eugenics, who hoped that genetics would provide the scientific basis for the postulated biological inequalities between people. In addition it was hoped that genetics would create the technical means to intervene in the process of inheritance (Müller-Hill 1991: 138; Weindling 1989: 430; Weingart 1987: 171, 183–90; Weingart/Kroll/Bayertz 1992: 320).

Two main schools of thought existed within eugenics, one focusing on positive selection, another on negative selection. Positive selection aimed at improving living conditions by means of social and health policies, which would create better circumstances for reproduction. By contrast, negative selection would prevent the reproduction of all those regarded as inferior. Proposals varied from controlling marriages to compulsory sterilization (Ganssmüller 1987: 12–16; Ludmerer 1969: 338; Weingart 1985: 318–20).

Oskar Vogt was among those scientists expressing support for eugenic measures, although he gave no clear indication whether he favoured positive or negative selection. His demands for the 'suppression of the inferior' (*Unterdrückung des Minderwertigen*) and the 'improvement of the superior' (*Höherzüchtung des Vollwertigen*) indicate that he favoured both methods. He was thereby convinced that one day brain research could provide the means of recognizing superior and inferior propensities of individuals based on brain anatomy. He argued that this created a scientific foundation for an analysis of individuals, the results of which exceeded those gained by psychological methods (Vogt 1912: 312–13; Vogt 1930b: 117; Vogt 1932).[10] In 1933, his wife Cécile wrote in the journal *Die Naturwissenschaften*:

> Our final aim is to comprehend mental and psychological characteristics (*die seelische Persönlichkeit*) and the possibilities of their inheritance (*ihrer Vererbungspotenzen*) for the furtherance of socially beneficial and the inhibition of negative propensities. This practical aim requires a prognosis for the individual and insights into possibilities for changing the predicted psychological and mental development (*der vorausgesehenen seelischen Lebensäußerungen*). (Vogt, Cécile 1933: 408).

According to Oskar Vogt, science should have a decisive influence on society as a result of its success in revealing the functional principles of nature. A modern state should therefore follow the recommendations of science in policy and legislation, as had already been demanded by Ernst Haeckel. From Vogt's point of view brain research should therefore be recognized, like the ideas of Haeckel, as a leading sector of science because it offered the possibility that one day hereditary diseases could be prevented, individuals inclined to commit crimes could be recognized early enough, and human society could be advanced by an improvement of the brain (Vogt 1912: 309–10; Vogt 1931: 309; Vogt, Cécile and Oskar 1919b: 245; Weß 1989: 13).

Oskar Vogt did not change his outlook when National Socialism first started to put eugenic ideas into practice, as occurred in July 1933 with the passing of the Law for the Prevention of Hereditary Diseased Offspring (Gesetz zur Verhütung erbkranken Nachwuchses). As a result of this law, about 350,000 people suffering from various disorders were sterilized between 1934 and 1945 (Degkwitz 1985: 215; Ganssmüller 1987: 34–50; Schmuhl 1987: 154–61; Weiss 1990: 43–4). The passing of this law was the outcome of a debate which had already begun before the First World War in many countries. Its text was based on a draft for a Sterilization Law written in 1932 by the Prussian Health Council. The main difference between the Prussian draft and the July 1933 law was that the latter comprised compulsory instead of voluntary sterilization of all those suffering from hereditary diseases. The Prussian draft still remained within the mainstream of eugenic legislation in various countries. For instance in 1920, compulsory sterilization for the mentally ill was legal in twenty-five countries. In the United States, the sterilization of the mentally ill had been practised since 1907. The final version of the 1933 law had abandoned this mainstream, but was still not totally unique. In Indiana, USA, the mandatory sterilization of feeble-minded patients had been introduced by law already before the First World War (Blasius 1991: 270; Degkwitz 1985: 214–15; Ganssmüller 1987: 18; Nachtsheim 1962: 1640–1; Weiss 1990: 39–40; Weß 1989: 16).

The Vogts were working on one of the hereditary diseases which fell under the Law for the Prevention of Hereditary Diseased Offspring, a disorder of the nervous system called Huntington's chorea or St Vitus's dance. In 1936, they pointed out that their research in this field provided the chance, 'to gain the eugenically important possibility to recognize before puberty all those with a hereditary disposition' for this disease (Vogt, Cécile and Oskar 1936: 396). They did not mention what should be done with these children; in fact they did not need to because this was conveniently determined by legislation.

Symptomatic of Oskar Vogt's conduct was the case of Max Bielschowsky, a distinguished neuropathologist, who had been working at the institute since 1904 (Kaminski 1990; Weil 1970). Archival sources show that in the late 1920s tensions arose between the two men.[11] Because Bielschowsky was Jewish he fell under the Law for the Reconstitution of the Professional Civil Service (Gesetz zur Wiederherstellung des Berufsbeamtentums) passed on 7 April 1933. Only two weeks later, Vogt informed the General Administration of the Kaiser Wilhelm Society that he intended to dismiss Bielschowsky. He was told that he was not permitted to do this without authorization by the Society's President.[12] Vogt reluctantly agreed to postpone the dismissal of Bielschowsky until the President was back in Berlin. Only three days earlier Vogt had sent a letter to the Society reminding that when Bielschowsky was appointed head of a department at the brain research institute by the administration of the Kaiser Wilhelm Society in 1925, he had already raised his objections. These had been rejected by the Society and, as he complained in his letter, were described then as 'expressions of an absolutely outdated anti-Semitism'.[13] Another document from the early 1920s concerning the cooperation between his institute and a Berlin hospital also indicates that Vogt already exhibited an anti-Semitic stance well before 1933. One of his accusations in a conflict with the director of this clinic was that the latter did not stick to the promise given to Vogt to recruit half of his employees from 'Aryan circles' (*aus arischen Kreisen*).[14] Although there is no definite evidence that Vogt used the Law for the Reconstitution of the Professional Civil Service to get rid of Bielschowsky at the earliest possible moment, the remarkable timing of his demand and the impetuosity with which he pursued his aim strongly support this assumption.

Vogt believed that in order to improve the brain a detailed knowledge of the process of inheritance was required. This corresponded with his general interest in questions of evolution, which he studied in bumble-bees. At the same he wanted to find out why some neurological disorders varied in frequency and severity. He thought that inter-generational anatomical and physiological variations in the cerebral cortex, which in his view determined mental and psychological characteristics, followed the same principles as the modifications of bumble-bees' colour patterns (Friedrich-Freksa 1961: 12; Hassler 1959/60: 246; Haymaker 1951: 181, 194–5; Richter 1996: 380–1). To learn more about the rules of inheritance he established a Genetics Department at his institute in 1925, headed by a young Soviet geneticist, Nikolai W. Timoféeff-Ressovsky. The two had met when Vogt was in Moscow to examine Lenin's brain after the latter had died in 1924. This was part of Vogt's studies on 'elite brains' with which he tried to reveal in the brain the anatomical foundations of geniuses (Eichler 1982: 287–8; Richter 1996: 375–6).

Vogt was among those trying to initiate and strengthen German–Soviet relations in science after the Rapallo Treaty of 1922, which created the favourable preconditions for this. Soviet researchers stayed repeatedly at Vogt's institute while he himself went to the Soviet Union several times. Besides examining Lenin, together with several other German neurologists he also presented the results of his work undertaken in Berlin on various occasions. In 1925, Vogt was appointed a Member of the Soviet Academy of Science, which shows in what high regard he was held in the Soviet Union. At the same time Vogt's activities were seen in Germany as instrumental in intensifying scientific ties with the Soviet Union. He repeatedly reported to the leadership of the Kaiser Wilhelm Society and the Foreign Office, which supported Vogt in his ventures (Nötzold 1990: 781; Richter 1976: 387–91; Richter 1996: 377–80; Weindling 1992: 185–9).

In 1928, a research institute was established in Moscow for the purpose of studying Lenin's brain and Vogt became its first director.

The research programme of this insitute went far beyond an examination of Lenin's brain. Vogt attempted to coordinate the work at Moscow and Berlin by allocating different research fields to each institute. While Berlin focused on normal and pathological anatomy and physiology of the brain, the Moscow institute concentrated on elite brain research. Vogt was also interested in the brain-anatomical differences between human races. To explore this he set up a department for race biology at the Moscow institute, headed by a staff member of the Berlin institute (Kirsche 1986: 13; Richter 1976: 389–99; Richter 1996: 377–80). Vogt described the purpose of this research as follows:

> The examination of brains from different races will help to answer the question of how far the cultural backwardness of some races is the result of milieu or of constitutional factors and which brain functions are developed best in these races. I do not have to go into further explanations about the importance of these results in the field of educational theory. (Vogt 1930a: 116)

Vogt headed the institute in Moscow until 1930. Then he decided to focus exclusively on his work in Berlin, where the new buildings of his institute were inaugurated on 2 June 1931 (Kirsche 1986: 11–15, 23; Richter 1996: 376–80).

Nikolai W. Timoféeff-Ressovsky

Timoféeff-Ressovsky was born in 1900 in Moscow where he studied biology in the early 1920s with Chetverikov, one of the leading Soviet geneticists. In 1925, Timoféeff accepted Vogt's offer to come to Berlin, where he initially continued his cross-breeding experiments with a type of fly called drosophila melanogaster. A US geneticist, Herman J. Muller, had discovered in the late 1920s that X-rays could be used to trigger mutations in drosophila, for which he was rewarded the Nobel Prize in 1946. This discovery opened the way to a new research field which became known as radiation biology. After the publication of Muller's results, Timoféeff started experimenting with X-rays and during the 1930s and 1940s became one of Germany's leading radiation biologists (Carlson 1981: 135–50; Dorna 1995: 16–19, 38–55; Paul/Krimbas 1992: 86).

Timoféeff's most important contribution to the advancement of science was the so-called 'target theory', which led the way to the

Figure 11   Nikolai W. Timoféeff-Ressovsky (1900–1981).

discovery of the nature of the gene and the emergence of molecular biology. He published the paper presenting target theory in 1935 with one of his collaborators, Karl-Günter Zimmer, and Max Delbrück, an assistant to Lise Meitner at the Kaiser Wilhelm Institute for Chemistry (Fischer 1988: 53–4, 78–82, 130–58, 230–4; Paul/Krimbas 1992: 86).

Like Vogt, Timoféeff was very interested in questions of evolution, which prompted him to pay special attention to mutations. These were particularly high in drosophila populations bred in laboratories compared to those in nature, leading to a serious deterioration of the former's genetic pool (Macrakis 1993: 519, 534; Paul/Krimbas 1992: 86–8; Roth 1986: 40–3, 51–2, 61–4). In 1935, Timoféeff published an article in *Der Erbarzt*, one of Germany's leading journals of eugenics, edited by Otmar Freiherr von Verschuer, director of the Kaiser Wilhelm Institute for Anthropology, Human Heredity and Eugenics from 1942 onwards. Transferring the results stemming from research on drosophila to humans, he argued that the suspension of the law of natural selection in civilized conditions favoured the spread of mutations. Assuming that this led to an increase in hereditary diseases, he demanded eugenic measures, although not mentioning which ones exactly, in order to prevent the deterioration of humanity's genetic pool (Timoféeff-Ressovsky 1935: 117).

Timoféeff was not alone in his beliefs. On the contrary, most geneticists assumed that inferior traits spread particularly fast in civilization, leading to a successive worsening of the genetic disposition of mankind. This caused, with striking similarity to the degeneration debate, an atmosphere of crisis among geneticists all over the world. In the United States for instance, approximately 50 per cent of all geneticists became involved in the eugenics movement before the First World War. And even until the late 1940s and early 1950s, various American and British geneticists remained committed to eugenics. To ensure that only individuals with first-rate genes could reproduce, geneticists in many countries demanded eugenic measures (Adams 1990: 219; Ludmerer 1969: 339–41, 345–6; Roth 1986: 35–6).

One of these was the above-mentioned Herman J. Muller, who was appointed a member of the Buch institute's advisory board in 1932.[15] Muller knew Vogt and it appears that they both shared the same outlook on eugenic questions. In 1932, Muller was invited to come to Buch where he spent some weeks before going to the Soviet Union. Already before the First World War Muller put forward eugenic ideas and pursued these aims until his death in 1967. In August 1939, at the 7th International Genetics Congress, Muller worked out a resolution which was signed by numerous renowned geneticists and biologists. It became known as the 'Geneticists' Manifesto' and demanded eugenic measures to prevent a further deterioration of humanity's genetic pool. Obviously none of those signing the text, and especially not Muller, were disturbed by the fact that such a policy had already been put into practice in Germany. In 1949, Muller in his function as president of the US Genetics Society repeated his concerns about the genetic deterioration of mankind and again demanded measures to prevent this (Roth 1985: 125–6, 129; Weß 1989: 108–54).

Due to the instruments genetic research promised for intervening into the mechanism of inheritance, NSDAP functionaries regarded genetics as essential for race hygiene (Müller-Hill 1991: 138–9). Oskar Vogt mentioned in his annual report in 1934 that 'some colleagues, mainly old party members' had asked Timoféeff for an introduction into the methods of experimental genetics. Timoféeff's research also helped Hermann Boehm of the Institute for Inheritance Studies Alt-Rehse (Institut für Vererbungslehre Alt-Rehse) to become familiar with this field.[16] This institute was part of the Leader School for German Doctors (Führerschule der deutschen Ärzteschaft) where physicians were instructed in (among other things) genetics and race hygiene (Kudlien 1985: 122–30).

This research field gained remarkable importance during the Third Reich. Experimental mutation research and radiation biology were among the most strongly supported fields of biological research. The increase of funds allocated by the German Research Foundation (Deutsche Forschungsgemeinschaft, DFG) and Reich Research Council (Reichsforschungsrat, RFR) far exceeded those of other subdivisions of botany and zoology (see tables 2 and 3, pp. 224–5).

In the spring of 1939, the Genetics Department put the first particle accelerator in Germany into operation to produce radioactive isotopes for experimental mutation research. Timoféeff quickly recognized that radioactive isotopes offered new possibilities for research. One

Table 2   Funding of Botany by DFG/RFR, 1934–1945

Fields	1934–9 RM	1940–5 RM	1934–9 Percentage	1940–5 Percentage
Morphology	20,950	19,740	2.7	2.1
Physiology	143,575	98,100	18.3	10.3
Genetics, Virology	100,845	181,993	12.9	19.1
Mutation, Radiation	55,224	234,522	7.0	24.6
Ecology	33,170	45,400	4.2	4.8
Medicinal Herbs	36,585	40,870	4.7	4.3
Applied Botany	393,668	331,248	50.2	34.8
Sum	784,017	951,873	100.0	100.0

Source: Deichmann 1992: 73.

Table 3 Funding of Zoology by DFG/RFR, 1934–1945

Fields	1934–9 RM	1940–5 RM	1934–9 Percentage	1940–5 Percentage
Morphology	18,400	0	3.6	0.0
Physiology	62,545	90,555	12.2	6.1
Developmental Physiology	38,670	26,640	7.5	1.8
Genetics, Virology	115,711	244,273	22.5	16.4
Mutation and Radiation	93,854	655,728	18.2	43.9
Cancer	75,971	178,420	14.8	12.0
Ecology	28,090	113,150	5.5	7.6
Applied Zoology	81,240	183,790	15.8	12.3
Sum	514,481	1,492,556	100.0	100.0

Source: Deichmann 1992: 80.

of these possibilities was the so-called indicator method, where radioactive isotopes were added to gas, and its direction could then be followed by registering the isotopes flowing with it (Born/Zimmer 1940a und 1940b; Born/Timoféeff-Ressovsky/Zimmer 1942; Richter 1996: 393–4). Timoféeff claimed that this method could be applied to various fields like testing if gas mask filters or cockpits of airplanes were hermetically sealed. To improve the indicator method he applied for financial support to the German Research Foundation, the Reich Research Council, various ministries and industry. An important partner in this and other fields was Auer Gesellschaft, a company involved in, among other things, the German nuclear bomb project.[17]

After the start of the Second World War radiobiological studies on the effects of various types of radiation on biological subjects were extended to fields relevant for the military. This included research on the effects of radiation on pilots when flying at very high altitudes and on the effects of nuclear radiation on animals and humans. Radioactive substances were also used as tracers for medical purposes. This was done by injecting Thorium-X into animals and humans to examine the circulation of the blood and to diagnose disturbances (Berg 1990: 464–5; Roth 1986: 34–5).[18] Archival sources also mention collaborative work starting in 1942 with the Army Research Centre (Heeresversuchsanstalt) at Peenemünde. Unfortunately no information could be found in the Bundesarchiv-Militärarchiv because the relevant documents no longer exist.[19] The research undertaken at the Genetics Department was granted high priority and was continued until the very end of the war. Between 1943 and 1945 some of the Department's work was financed by the Plenipotentiary of the Reichsmarschall for Nuclear Physics (Bevollmächtigter des Reichsmarschalls für Kernphysik), who organized research on nuclear power, including its military application. Even when work in this field was restricted in January 1945 to six groups, funding of the Buch Genetics Department continued (Deichmann 1992: 166–8).

Timoféeff and his department remained in Berlin while all others had moved to locations in the western parts of Germany from late 1944 onwards. Initially the Red Army allowed him to proceed with his research, but in July 1945 he was arrested by the Soviet Secret Service and taken to a prison camp in Kazakhstan. He only survived because his knowledge was useful for the ongoing nuclear weapons project. Therefore he was imprisoned in a military research centre and released only in 1955 during the reforms following Stalin's death. He then worked in various Soviet biomedical research institutes and died in 1981, still regarded as a traitor to the Soviet cause. His rehabilitation only happened after the dissolution of the Soviet Union (Eichler 1982; Glass 1990: 459–61; Paul/Krimbas 1992: 90–1).

There is no consensus in historical literature on how to judge Timoféeff. Disagreements focus on whether his research was relevant for armaments and if he contributed to the policy of race hygiene. Müller-Hill and Paul and Krimbas criticized the fact that he had stayed in Germany despite being offered a position in the United States in 1937. Roth accused Timoféeff of having transferred the results of his drosophila research to humans, thereby providing National Socialist race policy with science-based arguments (Müller-Hill 1988: 721–2; Paul/Krimbas 1992: 92; Roth 1986: 37).

The defendants of Timoféeff argued that he had made only limited concessions to the National Socialists, which essentially consisted of a few public statements on race hygiene and the collaboration with Hermann Boehm. Deichmann and Dorna reasoned that geneticists like Timoféeff only emphasized the importance of their research on race hygiene to receive sufficient financial support. Berg pointed to the fact that Timoféeff was a scientist of high personal integrity who had protected prisoners of war by employing them in his department, thereby risking his own life. Others also mentioned that one of his sons died in a concentration camp after being arrested for alleged contacts to a resistance group (Deichmann 1992: 169; Dorna 1995: 23; Berg 1990: 460, 466).

On the basis of the facts described above there can be hardly any doubt that the work undertaken at the Genetics Department, led by Timoféeff, was of military relevance. Even if this was not the case, Timoféeff should have realized that the Soviets would regard this research in Germany as being of military significance. Timoféeff's attitude towards eugenics is somewhat more problematic. Although there is enough evidence to show that he rejected National Socialist ideas in general, a careful study of his publications shows that he did support eugenic measures, not out of opportunism but on the assumption that mankind was jeopardized by a deterioration of its genetic pool leading to a faster spread of hereditary diseases. Contrary to many of his colleagues in the international genetics community, he did not mention which practical measures he had in mind but only pointed out the facts drawn from his experimental work. But again, as has been said in connection with Vogt, there was no need to be explicit about these measures because since 1933 they had already been put into practice.

Julius Hallervorden

Julius Hallervorden was born in 1882. He studied medicine in Königsberg and then worked at a psychiatric hospital in Brandenburg. Between 1927 and 1945 Hallervorden was the head pathologist of all state-run psychiatric hospitals in Brandenburg. In 1937, he joined the NSDAP and at the beginning of 1938 became head of the Department of Neurohistopathology at the Kaiser Wilhelm Institute for Brain Research. Early in 1945 this department moved to Hesse and in 1949 was incorporated into the newly established Max Planck Institute for Brain Research. The latter's director was the above-mentioned Hugo Spatz. Hallervorden worked at this institute until retirement, became the first president of the newly founded Union of German Neuropathologists (Vereinigung Deutscher Neuropathologen) and died in 1965 (Richardson 1990: 507–8, 510–11; Spaar 1966; Spatz 1966: 477–8).

Unlike the other two scientists presented here, Hallervorden concentrated not on basic research but, because he was trained as neuropathologist and came from clinical work, on various diseases of the nervous system. He dealt with multiple sclerosis and mental

Figure 12   Julius Hallervorden (1882–1965).

retardation, and achieved important results in child neurology, especially in the fields of neonatal neurological disorders and hereditary neurological diseases. This research required close contact with clinical medicine and the examination of anatomical material gained from post-mortem autopsies. To examine the nature, causes and development of particular disorders it was necessary to find a sufficient number of cases, preferably in various stages of the diseases. In this way similarities could be discovered and generalizations be made, opening the way to an earlier diagnosis and, in the long term, to successful treatment (Richardson 1990: 506–7, 510–11; Spatz 1966: 478–81).

The crucial point in the work of Hallervorden at the Kaiser Wilhelm Institute for Brain Research was connected with the so-called 'euthanasia-action T4' during which about 150,000 patients in psychiatric hospitals were killed. Among them were more than 5,000 children and adolescents, with whom the mass murder started in October 1939. The killings were undertaken in thirty psychiatric hospitals, Görden in Brandenburg being one of them. Görden was a special case in so far as it was the clinic where child euthanasia started. It was at the same time the only institution in the Reich where doctors were instructed in the killing of children by tablets and injections (Burleigh 1994: 100–1, 109; Friedlander 1995: 46–9; Klee 1985: 379–80, 458). As Brandenburg's head pathologist, Hallervorden was responsible for the autopsy of all those killed in this province, including Görden. In the course of his research he used the brains of more than 700 patients who had been murdered (Klee 1985: 395–6; Peiffer 1997: 34–9; Shevell 1992: 2216).

In June 1945, Hallervorden was questioned by Leo Alexander, a medical officer of the US Army. Alexander was a Jewish emigrant from Germany who had worked there with Oskar Vogt among others. When asked about the killings of patients Hallervorden answered, according to Alexander, that after he had heard about this he had asked for all those brains useful for his research, arguing that these would otherwise have been lost for science.[20] Hallervorden asserted later that he had not regarded the conversation with Alexander as an interrogation but as a talk among colleagues. He also denied having asked for brains and wrote to the Nuremberg International Military Tribunal requesting an official inquiry, which was rejected.[21]

Two authors dealing with this subject, Aly and Shevell, argued that the statement of Hallervorden as reported by Alexander had to be regarded as true. Shevell stated, for instance: 'It is clear that Hallervorden himself initiated the collaboration with those operating the euthanasia centers' (Shevell 1992: 2216; Aly 1994: 223–4). The fact that no one else was present makes one doubt that Alexander's report could be used to accuse Hallervorden. Furthermore, it appears more plausible that the initiative for using anatomical material stemming from patients killed at Görden and other psychiatric hospitals did not originate with Hallervorden. One of the doctors performing these killings, Heinrich Bunke, declared in his testimony at the Frankfurt Court of Justice in 1962–3: 'I mentioned relatively early in conversations that I thought it irresponsible not to use this abundant anatomical material. I proposed to extract the brains from some corpses and give them to the Kaiser Wilhelm Institute for Brain Research (Professor Dr Hallervorden).' (Hohmann/Wieland 1996: 164). Bunke also stated that he had asked Hallervorden if he could be instructed in the methods of brain pathology. For this purpose he spent several weeks in mid-1941 at the Institute for Brain Research where he was taught how to extract and conserve brains. When working afterwards at various hospitals in Brandenburg he extracted the brains of all those murdered patients whom he regarded of interest for further examination by Hallervorden and his department (Hohmann/Wieland 1996: 164).

This statement of Bunke suggests that the initiative to use the brains did not come from Hallervorden. But at the same time it must be emphasized that Hallervorden obviously utilized this anatomical material without any ethical considerations. There is no reason to assume that rejecting this material would have had any dire consequences, as was later asserted by Spatz, who argued that not accepting this material would have put the institute's existence at risk.[22] Normally Hallervorden sent one of his assistants to Görden to extract the brains while some of the autopsies in hospitals were performed by Hallervorden himself. On 8 March 1944 he informed Hermann Paul Nitsche, one of the organizers of the euthanasia programme: 'I received 697 brains altogether, including those which I extracted in Brandenburg myself.' Already in December 1942, Hallervorden had reported to the German Research Foundation (DFG) that his collection of neuropathological material 'was being completed by Brandenburgische Landesanstalten in Görden' (Müller-Hill 1984: 68).

We can only speculate as to how far Julius Hallervorden was involved in the process of selecting which patients to kill. When instructing Heinrich Bunke on how to extract and conserve brains it can be assumed that Hallervorden told him which diseases, and therefore which type of patient, were of special interest to him. Hallervorden, focusing especially on child neurology, also benefited from the fact that Görden had a department for children and adolescents. It was this hospital where 'euthanasia' of children started under Hans Heinze, director of Görden from 1939. The latter was one of only three expert referees who decided which children to murder in German hospitals and asylums (Burleigh 1994: 100–1; Friedlander 1995: 46–7, 57; Klee 1985: 300; Peiffer 1997: 35).

In 1941, Heinze established a research department at Görden, concentrating on various neurological diseases in children and

adolescents. One of his main research fields, extrapyramidal disorders, was also an area Hallervorden was working on, and it is known
that the two collaborated closely. By 1939, Heinze had become
member of the advisory board of the Kaiser Wilhelm Institute for
Brain Research (Aly 1994: 220; Knaape 1989: 226–7; Peiffer 1997:
35; Spatz 1966: 478–81).

One has to admit that Hallervorden was certainly right when
asserting that these killings would have taken place anyway. Nothing
is known about Hallervorden's attitude towards these activities while
they were still going on – later accounts emphasize that he inwardly
rejected them,[23] but whether this was actually true will never be
known. Looking at psychiatrists and neurologists in general, it can
be recognized that this community underwent a process of radicalization from the late 1920s onwards. The background for this was
the sharp rise in the number of psychiatric patients in the 1920s while
the financial resources available for treatment declined during the
world economic crisis. Attempts at reforming the system of psychiatric
hospitals increasingly focused on ways to deal with insufficient
financial means. The changes brought about by the deteriorating
economic situation helped to popularize eugenic ideas within the
medical community. Compulsory sterilization was at the core of the
proposals discussed and was intended to prevent a further spread of
hereditary diseases in order to relieve the strain on psychiatric
hospitals. At the same time it was hoped that this would free up
resources for the treatment of patients who were still regarded as
curable. All those who were not should be detained in low-cost
hospitals and asylums to save money. It is important to note that
some of those who later became the organizers of the 'euthanasia
programme' initially promoted reforms of the psychiatric system. This
was for instance true for Hermann Paul Nitsche (Aly 1994: 158–60;
Blasius 1991: 269–71; Peiffer 1992: 221–2; Siemen 1991: 193–9).

The main question in the case of Hallervorden is how to judge
somebody who used an opportunity created by the criminal activities
of others. A leading US neurologist, Webb Haymaker, who was later
involved in this subject, claimed that to condemn Hallervorden
because he asked for these brains was unjust. He and other US
neurologists argued that Hallervorden had always remained within
the ethical boundaries of his profession.[24] One of them stated in 1953:
'Even if Hallervorden had expressed a wish to receive the brains for
examination, this should not be regarded as criminal or unethical. It
is only natural that any neuropathologist might desire to examine

the brains in cases of severe chronic psychosis, knowing that, under the law, the patients were bound to be put to death.'[25]

Conclusions

First, despite the differences in research fields, methods and professional formation it is apparent that the three scientists had used the opportunities which arose in the Third Reich independently of their personal attitudes towards National Socialism. They can therefore hardly be regarded as 'Believers in an Age of Heresy', or in other words, as scientists who restricted themselves to the 'pure' progress of science. The exception was Oskar Vogt, but only because he was not given a chance to convince the Nazis of the potential benefits of his research. His dismissal as director of the Kaiser Wilhelm Institute for Brain Research was from this perspective not the result of an anti-Nazi stance but only came about because he, like many others, did not understand the true nature of Nazi repression. Continuing to act as he had done during the 1920s, threatening, intimidating and bullying everyone who refused to give in to his demands, he collided with the National Socialists' comprehension of dominance and power. Vogt continued to rely on the support of Krupp von Bohlen und Halbach whenever he ran into conflicts, but was finally forced to realize that even Krupp was not able to protect him.

Secondly, the three scientists presented here regarded themselves as apolitical, as did most members of the scientific community. They saw themselves as exclusively focusing on the advancement of science in the interests of medical and scientific progress. Sufficient funds to improve the basic conditions of research were gained by applying to various institutions like the German Research Foundation and the Reich Research Council, but also to industry. At the same time scientists presented the results of their research while obviously ignoring the ensuing consequences. Some may have been irritated that their proposals were hijacked by the National Socialists; others may have been relieved that long-standing demands, such as that for compulsory sterilization, were finally being put into practice.

If we turn from historical analysis to moral judgement, the main issue is function, not intention. In other words, it may be that scientists did not intend to support National Socialism with their research. But in fact they increased the credibility and authority of the system by publishing or by giving talks outside the scientific community. Arguing like this, one also has to consider what options scientists have in a totalitarian system and how they can evade the 'entangling alliances' which result from their need for sponsorship.

It is important to note that the ideas and visions of the three scientists presented here, like the notion that individuals and mankind as a whole could be biologically improved by a variety of means, still remained in the scientific mainstream. There was of course one fundamental difference between Germany and all other countries: it was only there that some of these concepts were put into practice within the context of an ideologically motivated policy of mass murder. That these ideas were so easily adaptable to the most terrible purposes shows how dubious the underlying logic behind them was.

Thirdly, compared to the perspectives their research offered, the practical policies implemented under National Socialism must have appeared rather primitive to many scientists. This was for instance true for anthropometric studies undertaken to gain indications of character and talents by measuring the body. Vogt, on the contrary, hoped to provide scientifically based information on this by studying the brain's anatomy, which in his view was closely linked to the mental and social characteristics of individuals. This does not mean that promising advances in science were ignored during the Third Reich. On the contrary, modern sectors of biological research like those undertaken by Timoféeff enjoyed high growth rates of public funding. This sheds an interesting light on National Socialist science policy and shows that any historical analysis focusing too strongly on obscure fields like Aryan Physics is bound to neglect important aspects of scientific development.

Fourthly, the three researchers presented here show how highly qualified scientists exhibited an entirely wrong perception of their own situation. Vogt assumed after 1933 that he could continue acting the way he had done before. Hallervorden was unscrupulous about using brains of patients killed by euthanasia, including those of children. Timoféeff undertook research of military significance without recognizing that for the Soviet Union he thereby became a traitor. And he then even stayed in Berlin while his colleagues fled from the advancing Red Army. These misperceptions had rather different consequences. Hallervorden only had to deal with accusations, but was able to continue his work. Vogt was dismissed as director but received funds for a new institute from Krupp. Timoféeff suffered the most, and only the importance of his work for the Soviet nuclear weapons programme saved his life.

Lastly, the developments described above illustrate that particular attention should be given to what could be called 'initiatives from below'. Or, in other words, to the attitudes and proposals from scientists, who for various reasons set their research and the results achieved at the disposal of the state. Ideas like those recounted here were not restricted to Germany or the Third Reich, and even after 1945 they enjoyed a certain popularity. This was particularly true of the sterilization of individuals regarded as not worthy of having children. In 1962 for example, Francis Crick, who with two others was awarded the Nobel Prize for discovering the structure of DNA, said the following in a discussion on 'eugenics and genetics':

> I want to concentrate on one particular issue: do people have the right to have children at all? It would not be very difficult, as we gathered from Dr. Pincus, for a government to put something into our food so that nobody could have children. Then possibly – and this is hypothetical – they could provide another chemical that would reverse the effect of the first, and only people licensed to bear children would be given this second chemical. This isn't so wild that we need not discuss it. Is it the general feeling that people do have the right to have children? This is taken for granted because it is part of Christian ethics, but in terms of humanist ethics I do not see why people should have the right to have children. I think that if we can get across to people the idea that their children are not entirely their own business and that it is not a private matter, it would be an enormous step forward. (Wolstenholme 1963: 275)

Statements like this serve as a vivid indicator of the 'strong continuities' (Bud 1993: 168) which exist between the pre- and post-war ethical attitudes of scientists, continuities which deserve further investigation.

Notes

1. MPG-Archiv, I/1A, Nr. 536, Report of Vogt to the Kaiser Wilhelm Society's General Administration, 21 March 1933; Minutes of a conference concerning the Kaiser Wilhelm Institute for Brain Research, 22 September 1933; Report of Artur Kreutzer, SA-Buch, 20 October 1933; Report concerning the incidents at the Kaiser Wilhelm Institute for Brain Research on 21 June 1933; MPG-Archiv, I/1A, Nr. 1582, Meeting of the

Advisory Board of the Kaiser Wilhelm Institute for Brain Research, 12 May 1936.

2. MPG-Archiv, II/1A, Personalia Oskar und Cécile Vogt, Max Planck to Bernhard Rust, Minister of Science and Education, 27 September 1934; ibid., Note, signed Telschow, 10 January 1935; Krupp von Bohlen und Halbach to Oskar Vogt, 30 November 1935; I/1A, Nr. 1590, Meeting of the Advisory Board of the Kaiser Wilhelm Institute for Brain Research, 1 November 1937.

3. MPG-Archiv, I/1A, Nr. 1577, Oskar Vogt to Gustav Krupp von Bohlen und Halbach, 11 January 1927; ibid., Nr. 1579, Friedrich Glum to Gustav Krupp von Bohlen und Halbach, 5 September 1929.

4. Various letters showing these efforts can be found in MPG-Archiv, I/1A, Nr. 1605.

5. MPG-Archiv, I/1A, Nr. 1613, Friedrich Glum, KWS, General Director, to Alan Gregg, Rockefeller Foundation, Paris, 8 March 1929; Morena S. Thompson, Rockefeller Foundation, to Adolf von Harnack, KWS, Chairman, 24 May 1929.

6. MPG-Archiv, I/1A, Nr. 1594, Oskar Vogt to Friedrich Glum, KWS, General Director, 30 March 1929; Oskar Vogt to KWS, General Administration, 17 June 1929; 'Begründung der Höhe der Bausumme und des Etats für das Kaiser Wilhelm Institut für Hirnforschung', Oskar Vogt, 25 October 1929; ibid., Nr. 1607, Bauabrechnung; Oskar Vogt to Friedrich Glum, 4 June 1929; ibid., Nr. 1605, Note, signed Morsbach, 5 July 1929; MPG-Archiv, II/1A, Personalia Oskar und Cécile Vogt, Gustav Krupp von Bohlen und Halbach to Oskar Vogt, 7 September 1930; Gustav Krupp von Bohlen und Halbach to Max Planck, 7 September 1930.

7. MPG-Archiv, I/1A, Nr. 536, Oskar Vogt, 1. Mitteilung auf Veranlassung der Generalverwaltung der Kaiser Wilhelm Gesellschaft zur Förderung der Wissenschaften, 25 April 1933; Friedrich Glum to Oskar Vogt, 27 April 1944.

8. MPG-Archiv, I/1A, Nr. 1590, Report of the Director to the Advisory Board of the Kaiser Wilhelm Institute for Brain Research, 17 July 1934; ibid., Nr. 536, Report on public tours at the Institute.

9. MPG-Archiv, I/1A, Nr. 1580, Das KWI für Hirnforschung, December 1930, [Oskar Vogt].

10. MPG-Archiv, I/1A, Nr. 1578, Kaiser Wilhelm Institut für Hirnforschung, Oskar Vogt, [1927/8].

11. MPG-Archiv, I/1A, Nr. 1579, Friedrich Glum to Gustav Krupp von Bohlen und Halbach, 5 September 1929. See also ibid., II/1A, Personalia Max Bielschowsky.

12. MPG-Archiv, II/1A, Personalia Max Bielschowsky, Oskar Vogt to Kaiser Wilhelm Society, General Administration, 18 April 1933; ibid., I/1A, Nr. 536, Note, signed Telschow, 21 April 1933.

13. MPG-Archiv, II/1A, Personalia Max Bielschowsky, Statement Oskar Vogt, 18 April 1933.
14. MPG-Archiv, I/1A, Nr. 1587, Oskar Vogt to Friedrich Glum, Kaiser Wilhelm Society, General Secretary, 11 April 1921.
15. Archiv der Berlin-Brandenburger Akademie der Wissenschaften, Bestand Kaiser Wilhelm Gesellschaft, 39 Wahlen der Kuratoriumsmitglieder des KWI für Hirnforschung in Berlin-Buch, Kuratoriumsvorsitzender an Kuratoriumsmitglieder, 3 December 1931.
16. MPG-Archiv, I/1A, Nr. 1590, Report of the Director to the Advisory Board of the Kaiser Wilhelm Institute for Brain Research, 17 July 1934.
17. MPG-Archiv, I/1A, Nr. 1583, Notice on a conversation between Forstmann, Kaiser Wilhelm Society, General Administration, and Timoféeff-Ressovsky on 17 November 1939, 20 November 1939; Forstmann to Rau, 10 February 1940; ibid., Nr. 1606, Timoféeff-Ressovsky to Reichsamt für Wirtschaftsaufbau, 23 April 1941; ibid., Nr. 1584, Timoféeff-Ressovsky to Reichsamt für Wirtschaftsaufbau, 25 March 1942 and also 31 March 1943.
18. MPG-Archiv, I/1A, Nr. 1584, Bericht der Genetischen Abteilung an das Reichsamt für Wirtschaftsaufbau zu den durchgeführten Versuchen mit künstlich radioaktiven Isotopen, unterzeichnet Timoféeff, 23 March 1942. A similar report was submitted by Timoféeff on 31 March 1943.
19. Bundesarchiv Berlin, Stiftung Archiv der Parteien und Massenorganisationen der DDR, DY 30 SED, Büro Prof. Kurt Hager, 1981–1989, Report, 2 June 1988. Inquiries of the author at the Bundesarchiv-Militärarchiv, Freiburg/Breisgau.
20. Alexander, Leo, Neuropathology and Neurophysiology, including Electro-Encephalography, in Wartime Germany, London: CIOS 1945 (=Office of US Chief of Counsel for the Prosecution of Axis Criminality No. L-170).
21. MPG-Archiv, II/1A, Personalia Julius Hallervorden, Julius Hallervorden to the President of the International Military Tribunal, 2 February 1946; Julius Hallervorden to Hugo Spatz, 15 June 1945; Julius Hallervorden to Alice Gräfin Platen, no date; Hugo Spatz to Leo Alexander, 25 March 1953.
22. MPG-Archiv, II/1A, Personalia Julius Hallervorden, Hugo Spatz to Leo Alexander, 25 March 1953.
23. MPG-Archiv, II/1A, Personalia Julius Hallervorden, Julius Hallervorden to Dr Telemann, Ärztekammer, no date; Hugo Spatz to Leo Alexander, 25 March 1953.
24. MPG-Archiv, II/1A, Personalia Julius Hallervorden, Webb Haymaker, USA, Armed Forces Institute of Pathology, Washington D.C. to Hugo Spatz, 9 July 1953; Robert Wartenberg, University of California Hospital, to Almeida Lima, Secretary General of the Fifth International Neurological Congress, Lisbon, Portugal, 29 April 1953.

25. MPG-Archiv, II/1A, Personalia Julius Hallervorden, Robert Wartenberg, University of California Hospital, to Almeida Lima, Secretary General of the Fifth International Neurological Congress, Lisbon, Portugal, 29 April 1953.

References

Adams, Mark B. (1990), 'Towards a Comparative History of Eugenics', in Mark B. Adams (ed.), *The Wellborn Science: Eugenics in Germany, France, Brazil and Russia*, New York/Oxford: Oxford University Press, 217–31.

Aly, Götz (1994), 'Pure and Tainted Progress', in Götz Aly et al. (eds), *Cleansing the Fatherland. Nazi Medicine and Racial Hygiene*, Baltimore: Johns Hopkins University Press, 156–237.

Berg, Raissa L. (1990), 'In Defense of Timoféeff-Ressovsky', *Quarterly Review of Biology*, 65: 457–79.

Bielka, Heinz (1997), *Die Medizinisch-Biologischen Institute Berlin-Buch. Beiträge zur Geschichte*, Berlin: Springer.

Blasius, Dirk (1991), 'Die "Maskerade des Bösen". Psychiatrische Forschung in der NS-Zeit', in Norbert Frei (ed.), *Medizin und Gesundheitspolitik in der NS-Zeit*, München: Oldenbourg, 265–85.

Born, Hans J. and Zimmer, Karl G. (1940a), 'Anwendung radioaktiver Isotope bei Untersuchungen über die Filtration von Aerosolen', *Die Naturwissenschaften*, 28: 447.

—— (1940b), 'Untersuchungen an Schwebstoff-Filtern', *Die Gasmaske*, 1940, No. 20: 1–5.

Born, Hans J., Timoféeff-Ressovsky, N. W. and Zimmer, K. G. (1942), 'Biologische Anwendung des Zählrohres', *Die Naturwissenschaften*, 30: 600–3.

Brocke, Bernhard vom (1990), 'Die Kaiser-Wilhelm-Gesellschaft in der Weimarer Republik. Ausbau zu einer gesamtdeutschen Forschungsorganisation (1918–1933)', in Rudolf Vierhaus and Bernhard vom Brocke (eds), *Forschung im Spannungsfeld von Politik und Gesellschaft. Geschichte und Struktur der Kaiser-Wilhelm-/Max-Planck-Gesellschaft*, Stuttgart: Deutsche Verlags-Anstalt, 197–355.

Bud, Robert (1993), *The Uses of Life. A History of Biotechnology*, Cambridge: Cambridge University Press.

Burleigh, Michael (1994), *Death and Deliverance. 'Euthanasia' in Germany c. 1900–1945*, Cambridge: Cambridge University Press.

Carlson, Elof A. (1981), *Genes, Radiation, and Society: The Life and Works of H. J. Muller*, Ithaca: Cornell University Press.

Degkwitz, Rudolf (1985), 'Medizinisches Denken und Handeln im Nationalsozialismus', *Fortschritte der Neurologie und Psychiatrie*, 53: 212–25.

Deichmann, Ute (1992), *Biologen unter Hitler: Vertreibung, Karrieren, Forschung*, Frankfurt a. M.: Campus.

Dorna, Wigbert N. W. (1995), *Timoféeff-Ressovsky in Berlin-Buch 1925– 1945. Sein Beitrag zur Genetik und dessen Verhältnis zur national- sozialistischen Erblehre*, Münster PhD.

Eichler, Wolfdietrich (1982), 'Zum Gedenken an N. N. Timoféeff- Ressovsky', *Deutsche Entomologische Zeitschrift*, 29: 287–91.

Finger, Stanley (1994), *Origins of Neuroscience. A History of Explorations into Brain Function*, Oxford: Oxford University Press.

Fischer, Ernst Peter (1988), *Das Atom der Biologen. Max Delbrück und der Ursprung der Molekulargenetik*, München: Piper.

Friedlander, Henry (1995), *The Origins of Nazi Genocide. From Euthanasia to the Final Solution*, Chapel Hill, NC: University of North Carolina Press.

Friedrich-Freksa, Hans (1961), 'Genetik und biochemische Genetik in den Instituten der Kaiser-Wilhelm-Gesellschaft und der Max- Planck-Gesellschaft', *Die Naturwissenschaften*, 48: 10–22.

Ganssmüller, Christian (1987), *Die Erbgesundheitspolitik des Dritten Reiches: Planung, Durchführung und Durchsetzung*, Köln: Böhlau.

Glass, Bentley (1990), 'The Grim Heritage of Lysenkoism: Four Personal Accounts. In Defense of Timoféeff-Ressovsky', *Quarterly Review of Biology*, 65: 457–79.

Harrington, Anne (1987), *Medicine, Mind and the Double Brain. A Study in Nineteenth-Century Thought*, Princeton: Princeton University Press.

Hassler, Rolf (1959/60), 'Oskar Vogt zum Gedächtnis, 6.4.1870– 31.7.1959', *Archiv für Psychiatrie und Nervenkrankheiten* 200: 239–56.

Haymaker, Webb (1951), 'Cécile and Oskar Vogt: On the Occasion of her 75th and his 80th Birthday', *Neurology*, 1: 179–204.

Hohmann, Joachim S. and Wieland, Günther (1996), *MfS-Operativ- vorgang 'Teufel'. 'Euthanasie'-Arzt Otto Hebold vor Gericht*, Berlin: Metropol.

Holdorff, B. (1989), 'Neurologie und Rassismus unter dem National- sozialismus', in Achim Thom and Samuel Mitja Rapoport (eds), *Das Schicksal der Medizin im Faschismus: Auftrag und Verpflichtung zur Bewahrung von Humanismus und Frieden. Internationales wissenschaftliches Symposium europäischer Sektionen der IPPNW, 17.–20. November 1988, Erfurt/Weimar, DDR*, Neckarsulm/München: Jungjohann, 131–3.

Kaminski, Henry J. (1990), 'Max Bielschowsky', in Stephen Ashwal (ed.), *Founders of Child Neurology*, San Francisco: Norman Publ., 426–30.

Kirsche, Walter (1986), 'Oskar Vogt 1870–1959: Leben und Werk und dessen Beziehung zur Hirnforschung der Gegenwart', *Sitzungsberichte der Akademie der Wissenschaften der DDR: Mathematik, Naturwissenschaft, Technik* 1985, no. 13/N, Berlin: Akademie-Verlag.

Klee, Ernst (1985), *'Euthanasie' im NS-Staat: Die 'Vernichtung unwerten Lebens'*, Frankfurt a. M.: Fischer.

Knaape, H.-H. (1989), 'Die medizinische Forschung an geistig behinderten Kindern in Brandenburg-Görden in der Zeit des Faschismus', in Achim Thom and Samuel Mitja Rapoport (eds), *Das Schicksal der Medizin im Faschismus: Auftrag und Verpflichtung zur Bewahrung von Humanismus und Frieden. Internationales wissenschaftliches Symposium europäischer Sektionen der IPPNW, 17.–20. November 1988, Erfurt/Weimar, DDR*, Neckarsulm/München: Jungjohann, 224–7.

Kudlien, Fridolf (1985), *Ärzte im Nationalsozialismus*, Köln: Kiepenheuer & Witsch.

Ludmerer, Kenneth M. (1969), 'American Geneticists and the Eugenics Movement: 1905–1935', *Journal of the History of Biology*, 2: 337–62.

Macrakis, Kristie (1993), 'The Survival of Basic Biological Research in National Socialist Germany', *Journal of the History of Biology*, 26: 519–43.

Magner, Lois N. (1979), *A History of the Life Sciences*, New York/Basel: Dekker.

Mayr, Ernst (1984), *Die Entwicklung der biologischen Gedankenwelt. Vielfalt, Evolution und Vererbung*, Berlin: Springer.

Müller-Hill, Benno (1984), *Tödliche Wissenschaft: Die Aussonderung von Juden, Zigeunern und Geisteskranken 1933–1945*, Reinbek bei Hamburg: Rowohlt.

—— (1988), 'Heroes and Villains. A Review of Der Genetiker: Das Leben des Nikolai Timoféeff-Ressovsky genannt Ur, by D. Granin', *Nature*, 336: 721–2.

—— (1991), 'Selektion. Die Wissenschaft von der biologischen Auslese des Menschen durch Menschen', in Norbert Frei (ed.), *Medizin und Gesundheitspolitik in der NS-Zeit*, München: Oldenbourg, 137–55.

Nachtsheim, Hans (1962), 'Das Gesetz zur Verhütung erbkranken Nachwuchses aus dem Jahre 1933 in heutiger Sicht', *Ärztliche Mitteilungen*, 47/59: 1640–4.

Nötzold, Jürgen (1990), 'Die deutsch-sowjetischen Wissenschafts-beziehungen', in Rudolf Vierhaus and Bernhard vom Brocke (eds), *Forschung im Spannungsfeld von Politik und Gesellschaft. Geschichte und Struktur der Kaiser-Wilhelm-/Max-Planck-Gesellschaft*, Stuttgart: Deutsche Verlags-Anstalt, 778–801.

Paul, Diane B./Krimbas, Costas B. (1992), 'Nikolai W. Timofejew-Ressovski', *Scientific American*, 2: 86–92.

Peiffer, Jürgen (1992), 'Damals und Heute. Ethische Konfliktsituation des wissenschaftlich arbeitenden Arztes', in Jürgen Peiffer (ed.), *Menschenverachtung und Opportunismus. Zur Medizin im Dritten Reich*, Tübingen: Attempto, 213–44.

―― (1997), *Hirnforschung im Zwielicht: Beispiele verführbarer Wissenschaft aus der Zeit des Nationalsozialismus*, Husum: Matthiesen.

Pick, Daniel (1989), *Faces of Degeneration. A European Disorder, c. 1848–c. 1918*, Cambridge: Cambridge University Press.

Porter, Theodore M. (1990), 'Natural Science and Social Theory', in Robert Olby et al. (eds), *Companion to the History of Modern Science*, London: Routledge, 1024–43.

Richardson, Edward P. (1990), 'Julius Hallervorden', in Stephen Ashwal (ed.), *Founders of Child Neurology*, San Francisco: Norman, 508–12.

Richter, Jochen (1976), 'Oskar Vogt, der Begründer des Moskauer Staatsinstituts für Hirnforschung', *Psychiatrie, Neurologie und medizinische Psychologie*, 28: 385–95.

―― (1996), 'Das Kaiser-Wilhelm-Institut für Hirnforschung und die Topographie der Großhirnhemisphäre', in Bernhard vom Brocke and Hubert Laitko (eds), *Die Kaiser-Wilhelm-Gesellschaft / Max-Planck-Gesellschaft und ihre Institute. Studien zu ihrer Geschichte: Das Harnack-Prinzip*, Berlin/New York: de Gruyter, 349–408.

Roth, Karl Heinz (1985), 'Sozialer Fortschritt durch Menschen-züchtung: Der Genetiker und Eugeniker H. J. Muller (1890–1967)', in Friedrich Hansen and Regine Kollek (eds), *Gentechnologie: Die neue soziale Waffe*, Hamburg: Konkret Literatur-Verlag, 120–51.

―― (1986), 'Schöner neuer Mensch – Der Paradigmenwechsel in der klassischen Genetik und seine Auswirkungen auf die Bevölke-rungsbiologie des "Dritten Reichs"', in Heidrun Kaupen-Haas (ed.), *Der Griff nach der Bevölkerung. Aktualität und Kontinuität nazistischer Bevölkerungspolitik*, Hamburg: Delphi, 11–63.

Schmuhl, Hans-Walter (1987), *Rassenhygiene, Nationalsozialismus, Euthanasie. Von der Verhütung zur Vernichtung 'lebensunwerten Lebens', 1890–1945* (Kritische Studien zur Geschichtswissenschaft 75), Göttingen: Vandenhoeck & Ruprecht.

Shevell, Michael (1992), 'Racial Hygiene, Active Euthanasia, and Julius Hallervorden', *Neurology*, 42: 2214–19.

Siemen, Hans Ludwig (1991), 'Reform und Radikalisierung. Veränderung der Psychiatrie in der Weltwirtschaftskrise', in Norbert Frei (ed.), *Medizin und Gesundheitspolitik in der NS-Zeit*, München: Oldenbourg, 191–200.

Spaar, F. W. (1966), 'Julius Hallervorden – 1882 bis 1965', *Elektromedizin*, 11: 109–10.

Spatz, Hugo (1966), 'Erinnerungen an Julius Hallervorden (1882–1965)', *Der Nervenarzt*, 37: 477–82.

Timoféeff-Ressovsky, Nikolai (1935), 'Experimentelle Untersuchungen der erblichen Belastung von Populationen', *Der Erbarzt*, 8: 117.

Vogt, Cécile (1933), 'Warum stellen wir die Hirnforschung in den Mittelpunkt unserer Forschung?', *Die Naturwissenschaften*, 21: 408–10.

Vogt, Oskar (1912), 'Bedeutung, Ziele und Wege der Hirnforschung', *Nord und Süd*, 36: 309–14.

—— (1930a), 'Das Kaiser Wilhelm-Institut für Hirnforschung', in Ludolph Brauer et al. (eds), *Forschungsinstitute. Ihre Geschichte, Organisation und Ziele*, Hamburg: Hartung, vol. 2: 116–21.

—— (1930b), 'Erster Bericht über die Arbeiten des Moskauer Staatsinstituts für Hirnforschung', *Journal für Psychologie und Neurologie*, 40: 116.

—— (1931), 'Warum treiben wir Hirnforschung', *Forschungen und Fortschritte der deutschen Wissenschaft*, 7: 309.

—— (1932), 'Neurology and Eugenics', *Eugenics Review*, 24, no. 1: 15–18.

Vogt, Cécile and Vogt, Oskar (1919a), 'Allgemeinere Ergebnisse unserer Hirnforschung, 1.–4. Mitteilung', *Journal für Psychologie und Neurologie*, 25, Ergänzungsheft 1: 279–462.

—— (1919b), 'Wissenschaftliche Forderungen an den modernen Staat', *Nord und Süd*, 43: 245–50.

—— (1936), 'Kaiser Wilhelm-Institut für Hirnforschung in Berlin-Buch', in *25 Jahre Kaiser Wilhelm-Gesellschaft zur Förderung der Wissenschaften*, Berlin: Springer, vol. 2: 387–400.

Weber, Matthias M. (1991), 'Psychiatrie als "Rassenhygiene". Ernst Rüdin und die Deutsche Forschungsanstalt für Psychiatrie in München', *Medizin in Geschichte und Gesellschaft*, 10: 149–69.

Weil, Arthur (1970), 'Max Bielschowsky (1869–1940)', in Webb Haymaker (ed.), *The Founders of Neurology*, Springfield, Ill.: Thomas, 319–22.

Weindling, Paul (1988), 'From Philanthropy to International Science Policy. Rockefeller Founding of Biomedical Sciences in Germany 1920–1940', in Nicolaas A. Rupke (ed.), *Science, Politics and the Public Good. Essays in Honour of Margret Gowing*, London: Macmillan, 119–40.

—— (1989), *Health, Race and German Politics between National Unification and Nazism, 1870–1945*, Cambridge: Cambridge University Press.

—— (1992), 'German-Soviet Medical Cooperation and the Institute for Racial Research, 1927–c.1935', *German History*, 10: 177–206.

Weingart, Peter (1985), 'Eugenik – Eine angewandte Wissenschaft. Utopien der Menschenzüchtung zwischen Wissenschaftsentwicklung und Politik', in Peter Lundgreen (ed.), *Wissenschaft im Dritten Reich*, Frankfurt a. M.: Suhrkamp, 314–49.

—— (1987), 'The Rationalization of Sexual Behaviour: The Institutionalization of Eugenic Thought in Germany', *Journal of the History of Biology*, 20: 159–93.

Weingart, Peter/Kroll, Jürgen/Bayertz, Kurt (1992), *Rasse, Blut und Gene. Geschichte der Eugenik und Rassenhygiene in Deutschland*, Frankfurt a. M.: Suhrkamp.

Weiss, Sheila Faith (1990), 'The Race Hygiene Movement in Germany, 1904–1945', in Mark B. Adams (ed.), *The Wellborn Science: Eugenics in Germany, France, Brazil and Russia*, New York/Oxford: Oxford University Press, 8–68.

Weß, Ludger (1989), *Die Träume der Genetik. Gentechnische Utopien von sozialem Fortschritt*, Nördlingen: Greno.

Wolstenholme, Gordon E. W. (ed.) (1963), *Man and his Future. A Ciba Foundation Volume*, London: Churchill.

Zmarzlik, Hans-Günter (1963), 'Der Sozialdarwinismus in Deutschland als geschichtliches Problem', *Vierteljahreshefte für Zeitgeschichte*, 11: 246–73.

UTE DEICHMANN

The Expulsion of German-Jewish Chemists and Biochemists and their Correspondence with Colleagues in Germany after 1945: The Impossibility of Normalization?

This study focuses on the dismissal from academia and the forced emigration of all Jewish chemists and biochemists from Nazi Germany and Austria and on the few outspokenly liberal or leftist non-Jews. In addition, this study includes an analysis of selected post-war correspondences of émigrés with their German colleagues. Notwithstanding certain American and British achievements in physical chemistry, Germany was the international leader in chemistry and biochemistry until the mid-1930s. Due to the high percentage of Jewish scientists, both fields were strongly affected by the expulsion of scientists from Germany that began in 1933. Organic and inorganic chemistry were least affected, biochemistry suffered most, and polymer chemistry and quantum chemistry, already of minor importance among the majority of academic chemists before 1933, were further weakened by the expulsion of renowned scientists.

In the first part of this study, I present some data on the dismissal and forced emigration of academic (bio-)chemists and analyse the reactions of those scientists who remained in Germany. This part is based on a quantitative and qualitative analysis of this topic.[1] The analysis of the post-war correspondence of exiled German-Jewish chemists and biochemists with their former colleagues in Germany (after 1949 the Federal Republic of Germany) constitutes the second and major part of the study.[2] It was supplemented by information I received in interviews which I conducted with émigré and non-émigré (bio)chemists. Publications dealing with post-war relations of émigré scientists and former colleagues in Germany indicate, on the one hand, that German scientists were reintegrated easily into the international scientific community, that friendships were renewed, and that émigrés were of great help to German colleagues (e.g. Walker 1990: 229; Nordwig 1983; Jaenicke 1989). On the other hand, it has been shown that the National Socialist past and the post-war attempts to deny or ignore it by prominent German scientists put a severe strain on previous friendships and contributed to preventing a renewal of scientific cooperation (e.g. Fölsing 1993: 816–17; Wise 1994: 251–3; Sime 1996: 334–46).

My analysis focuses on a variety of individual cases of mostly prominent scientists, who reveal the characteristic attitudes and sensitivities of the post-war era. I do not attempt to give a complete overview of this kind of correspondence throughout the discipline of (bio)chemistry, and I do not include chemists from industry. It must also not be forgotten that an unknown number of scientists in exile were reluctant to resume contact with Germans again, at least during the first five to ten post-war years. The fact that some of the most eminent and influential German (bio)chemists openly supported aspects of Nazi ideology and politics and therefore had difficulties in resuming international relations after 1945 demonstrates the impact of political and psychological factors on the post-war development of (bio)chemistry in Germany. By analysing the correspondence and the biographical and scientific background of the correspondents, I have attempted to answer the following questions: What were the main motives of German (bio)chemists in resuming contact with their colleagues in exile and vice versa? What were the attitudes of German (bio)chemists towards the National Socialist past, and did they face it openly and self-critically? To what extent did émigré (bio)chemists aid their German colleagues scientifically and help them to resume international contacts? Did the post-war correspondence indeed mark the beginning of a 'normalization'[3] of scientific exchange and personal communication after the twelve years of the Third Reich?

The Expulsion and Forced Emigration of Chemists and Biochemists from German Academia, 1933 to 1939, and Some Reactions of non-affected German Scientists

Data on the Dismissal and Emigration

As is well known, the purge of universities and Kaiser Wilhelm Institutes (KWIs) of Jews and scientists with left-wing sympathies began with the Law for the Reconstitution of the Professional Civil Service, effective on 7 April 1933. The dismissed 'non-Aryans' were persons who had at least one Jewish, i.e. non-baptized, grandparent. Exemptions granted to former Jewish front-line soldiers were abolished by the Nuremberg laws of 1935. Non-Jewish men who were married to a Jew were usually dismissed in 1937. German laws came into force at Austrian universities immediately after the *Anschluss* on 13 March 1938 and at the German University of Prague in 1939.

Of the 140 people expelled from their positions (26.1 per cent of all 535 (bio)chemists in 1933/8), at least 122 (87 per cent) were for 'racial' reasons, either because they were classified as 'non-Aryans' or because they had a Jewish wife. Some of those dismissed for 'racial' reasons were also so-called 'political undesirables'. Six individuals, that is, 4 per cent of those (bio)chemists who were removed from their positions, were dismissed solely for political reasons, which included people who openly disagreed with aspects of Nazi policy, as well as Catholics who distinguished themselves by an especially strong bond to the Pope. Three of them were reinstated after a short time. At least 108 (20.1 per cent of all (bio)chemists in 1933/8) emigrated. Only four of the émigrés 108 returned to positions at German universities after 1945 prior to retirement.

The dismissals and forced emigrations in (bio)chemistry, made up 26 per cent of all (bio)chemists, and represented about twice the percentage in biology where 13 per cent were dismissed and 10 per cent emigrated (Deichmann 1996: 25). The dismissal and emigration quota in (bio)chemistry seems to have been as large as in physics, where the available data suggests these were between 15.5 and 25 per cent dismissals (Beyerchen 1977, Fischer 1988). This can be related to the relatively high proportion of Jews in academic (bio)chemistry and physics, compared to biology. In chemistry, since the 1850s an increasing number of academic chemists received posts in industry, where anti-Semitism was often less of an impediment to Jews. Many Jewish students studied chemistry in order to embark on industrial careers; others remained as university teachers – usually in the lower position of an associate professor – or as researchers at Kaiser Wilhelm Institutes. The biochemist Erwin Chargaff, for example, studied chemistry after considering that it 'was the one science that most easily provided opportunities for employment, either in the field of teaching, or in research or industry'.[4]

A comparison of the number of citations of émigrés and non-émigrés in the Science Citation Index for 1945–54 shows that the losses for Germany were even more considerable than was indicated by the number of dismissals and emigrations: non-émigrés were cited on average 150 times, émigrés 360 times – that is, 2.4 times more than non-émigrés.[5] Different areas of (bio)chemistry were affected to a very different extent by the expulsions. If the Nobel Prize, the number of citations in the Science Citation Index of 1945–54, and the individual major contributions to (bio)chemistry are taken as a measure of excellence, then the émigré scientists of the following table are most eminent (see table 4, pp. 248–9).

The table demonstrates the large number of biochemists (some of them chemists who later worked predominantly in biochemistry or in the chemistry of natural products) among the twenty-five most distinguished dismissed and/or émigré chemists and biochemists. Most of the individuals listed in the table had positions or posts at German or Austrian universities or Kaiser Wilhelm Institutes before they were forced to emigrate. Erwin Chargaff, Fritz Lipmann, Leonor Michaelis, and Max Perutz left Germany and Austria before they were expelled. Except for Werner Kuhn all the scientists were Jewish or had Jewish forebears. Kuhn was a Swiss citizen who had lived in Germany since 1927; the increasing possibility of a European war played a central role in his decision to leave Germany. Except for Bloch and Perutz this table does not include distinguished (bio)-chemists who were forced to emigrate when they were still students. For an analysis of the impact of the dismissals and emigrations on various chemical subdisciplines and on the countries where émigrés found refuge see Deichmann (1999).

Reactions of non-affected German Scientists

It has often been asked: how was it that this purge of Jews – who in many cases strongly identified with the German nation and its culture – could take place without a public outcry from their non-Jewish colleagues? Obedience to the law, a widespread (though usually not openly expressed) anti-Semitism among German and Austrian academics, and opportunism explain most of this silent complicity and absence of solidarity. These reasons also apply to most of those scientists who did not join the Nazi party or rejected other aspects of National Socialism. Karl Freudenberg, a non-Jewish internationally renowned organic chemist and professor at Heidelberg University, showed that it was not fear of National Socialist repression which made him accept and justify the expulsion of his Jewish colleagues. In a letter to his British colleague George Barger, professor of chemistry at the University of Edinburgh, who helped many refugees by giving them work in his department, Freudenberg responded to Barger's criticism of the dismissals in Germany:[6]

> You ask why the I.G. [I.G. Farben] does not keep colleague Löwe. I believe that this company tries to do what is possible but I assume that in this respect it also has to follow certain orders which are compulsory for all public institutions. By this I also have to justify the mild reproach which you directed against the German Chemical Association. There are orders which you simply have to comply with. It is my firm conviction that a cure of the body of the German people (*Kur am Leibe des deutschen Volkes*) was necessary, something which probably only very few will deny. The way it has been carried out cannot be subject to lengthy considerations in this country, simply because there are orders, and it does not matter at all, what the viewpoint of an individual is. I understand as well that opinions differ (*Meinungen sind geteilt*) in foreign countries whose attitude is that of an observer. It is not to be seen as a rejection of our new German state if we show a grateful appreciation (*begrüßen*) of expressions of help for the individual, which can be found particularly in England.

Freudenberg regretted that other countries focused on the problem of anti-Semitism which, according to him, though important, was only a small part of what was happening, 'and I may say that with most educated (*gebildete*) Germans an appreciation (*Anerkennung*) of the achievements is prevalent. We have a united central government of great power, and an absolutely pure man at its head, who is simplifying the administration with great energy.'[6]

Freudenberg's reference to the curing of the German body is reminiscent of the concept of 'cleansing' that was so central to Nazi thought. There were, in the eyes of many, including academics, too many Jews and too many political parties. Hitler promised to solve

Table 4 Twenty-five dismissed and/or émigré (bio)chemists, who received the most citations in the Science Citation Index 1945–54 and/or a Nobel Prize or distinguished themselves by individual major scientific contributions. Abbreviations: U.: University; TH: Technical University; KWI: Kaiser Wilhelm Institute: Med. Res.: Medical Research; ass.: assistant; res. fel.: research fellow; Inst. Institute.

Name	Discipline	Nobel Prize	Institution in Germany/Austria	First position after emigration
Max Bergmann	chemistry/biochemistry		KWI Leather Research (director)	Rockefeller Inst. of Med. Res.
Konrad Bloch	chemistry/biochemistry	1964	TH Munich (student)	Columbia U.
Ernst Boris Chain	chemistry/biochemistry	1945	U. Berlin (res. fel.)	Oxford U
Erwin Chargaff	chemistry/biochemistry		U. Berlin (ass.)	Columbia U.
Gustav Embden	biochemistry		U. Frankfurt, died in 1933	
Felix Haurowitz	biochemistry		German U. Prague (ass. prof)	U. Istanbul
Hans Krebs	biochemistry	1953	U. Freiburg (ass.)	U. Sheffield
Fritz Lipmann	biochemistry/chemistry	1953	KWI Med. Res. (res. fel.)	Carlsberg, Biol. Inst., Copenhagen
Otto Loewi	pharmacology/biochemistry	1936	U. Graz (prof.)	Univ. College N.Y.
Otto Meyerhof	biochemistry	1922	KWI Med. Res. (director)	Inst. Biol. Phys. Chimique, Paris
Leonor Michaelis	biochemistry		U. Berlin (ass. prof.)	Rockefeller Inst. of Med. Res.
David Nachmansohn	biochemistry		Virchow Hospital Berlin (ass.)	Sorbonne

Name	Discipline	Nobel Prize	Institution in Germany/Austria	First position after emigration
Carl Neuberg	chemistry/biochemistry		KWI Biochem. (director)	
Max Perutz	chemistry/biochemistry	1962	U. Vienna (student)	Lab. Molec. Biol., Cambridge
Rudolf Schönheimer	biochemistry		U. Freiburg (lecturer)	Columbia U.
Richard Willstätter	chemistry/biochemistry	1915	U. Munich (prof. until 1924)	–
Fritz Arndt	organic chemistry		U. Prague	U. Istanbul
Ernst D. Bergmann	organic chemistry		U. Berlin (lecturer)	D. Sieff (Weizmann) Inst.
Rudolf Lemberg	organic chemistry		U. Heidelberg (lecturer)	Royal Hospital, Sidney
Fritz Haber	physical chemistry	1919	KWI Physical Chem. (director), U. Berlin (prof.)	–
George de Hevesy	physical chemistry	1943	U. Freiburg (prof.)	U. Copenhagen
Werner Kuhn	physical chemistry		U. Kiel (prof.)	U. Basle
Hermann F. Mark	physical chemistry		U. Vienna (prof.)	Industry, Canada
Otto Stern	physical chemistry	1943	U. Hamburg (prof.)	Carnegie Inst., Pittsburgh
Joseph J. Weiss	physical chemistry		KWI Physical Chem. (ass.)	U. Durham

the problems created by both. And Freudenberg, like most of his colleagues, was an obedient civil servant. Nevertheless, it should be mentioned that he had close connections with his Jewish colleagues, for example, Max Bergmann, the director of the KWI for Leather Research in Dresden, who was expelled in the summer of 1933. They corresponded extensively from 1919 until 1935,[7] and Freudenberg greatly appreciated Bergmann as a scientist and person (and vice versa). In 1940 Freudenberg wrote a favourable review of a textbook written by Hermann Mark, a former professor of physical chemistry at the University of Vienna who had been expelled as a 'non-Aryan' three years before (Freudenberg 1940). Freudenberg's daughter was married to a Jew. He was not a simple anti-Semite, and he never became a member of the Nazi party. But like so many of his colleagues he was not opposed in principle to the expulsion of his Jewish colleagues and even welcomed it as a necessary sacrifice or remedy for German problems.

An extended correspondence between the two relatively young non-Jewish physical chemists Paul Harteck (born 1902), a postdoctoral assistant (*Assistent*) of Fritz Haber at the KWI for Physical Chemistry in Berlin, and Karl-Friedrich Bonhoeffer (born 1899), at the time full professor at the University of Frankfurt,[8] characterizes the prevailing atmosphere during 1933, described by Freudenberg. Neither joined the Nazi party, both despised Nazi ideology, and Bonhoeffer (like some other professors) provided work conditions in his institute for 'half-Jews' during the first years of the Second World War. Though Bonhoeffer and Harteck remained friendly with some of their dismissed colleagues, they refused to express solidarity with them as a specific group of individuals that was being treated unjustly. Thus Harteck, who in 1933 spent a year as research fellow with Rutherford in Cambridge, wrote to Bonhoeffer on 16 April 1933: 'When you come to Berlin . . . don't let the former members of the institute [KWI for Physical Chemistry] talk you into anything. They tend to say "the decent Aryans, too, ought to sympathize".' On 5 May he wrote: 'In London the Jews, the half and quarter Jews of Germany are gathering. If these people have ever had a liking for Germany, it can only have been a very superficial one, because now you really don't notice anything of it.' Harteck did not seem to comprehend the feelings of his former compatriots.

The correspondence later in 1933 concentrated on Harteck's need to find a new position. By the end of 1933 it became clear that the non-Jewish assistants of Fritz Haber's KWI for Physical Chemistry would also not have their posts extended, because Haber's successor, the convinced National Socialist Gerhart Jander, brought in a new staff of his choice (Szöllösi-Janze 1998: 677–8). Thus the dismissal of the Jews was of importance to Harteck's career. Bonhoeffer, who informed Harteck about developments in Germany, wrote to him on 22 June 1933: 'There are many openings occurring simultaneously, you should actually get one of them.' Harteck filled one of the vacancies made available by the dismissal of Jews: in 1934 he became successor to Otto Stern as full professor of physical chemistry at Hamburg University.

In this respect, Paul Harteck's experience was typical of many young non-Jewish scientists who soon found themselves in more senior posts than would normally have been available to them. The replacement of mostly older professors by young ones who were often active National Socialists was regarded with approval by students. Against the background of unemployment among academics, the fact that many non-Jewish German scientists benefited from the expulsion of their Jewish colleagues was one of the most important reasons for the ready acceptance of the Nazi policy of expulsion at universities.

Opportunism and a nationalistic conviction were driving forces also for Richard Kuhn, an internationally leading chemist in vitamin research and Nobel laureate of 1938 (the Nazis did not, however, allow Germans to accept the Nobel Prize after 1935). Kuhn, since 1929 head of the department of chemistry of the KWI for Medical Research in Heidelberg, became an influential chemist in the Third Reich despite his not joining the Nazi party. Because he was an internationally renowned chemist, followed every order of the Nazi regime without question, and closely collaborated with high-ranking Nazi officials, he succeeded in receiving and maintaining a powerful position and very high funding (see below).[9] In 1936 he denounced his colleague, the biochemist Otto Meyerhof, to the General Administration of the KWG for giving laboratory space to Hermann Lehmann, a Jewish biochemist.[10] Meyerhof, a Nobel Prize winner (1922), was head of the department of physiology at this KWI. Lehmann had to leave the institute immediately and later emigrated to the UK. Meyerhof was Jewish himself but was allowed to stay on until 1938, when he was dismissed and emigrated, first to France and then, in 1940, to the United States, where he became a visiting professor at the University of Pennsylvania in Philadelphia. In 1937 Kuhn became director of the entire KWI for Medical Research. In 1938 he was appointed head of the Association of German Chemists and in 1940 also head of the

Chemistry Section of the Reich Research Council. He was thus entitled to decide about the allocation of research funds from government to organic chemists. Kuhn's address to the Association of German Chemists on the occasion of its 75th anniversary in 1942 exposes the German nationalistic conviction that made him an unconditional follower of Hitler. He ended his address by saying (Kuhn 1942):

> We realize the overwhelming contribution to the foundation of modern chemistry of those occidental people who donated to mankind a Scheele and Berzelius, a Lavoisier and Pasteur, a Liebig and a Wöhler. The peoples of Europe are under arms (*stehen die Völker Europas heute unter den Waffen*) for the continuation of this blood, and the further development of their culture in the same way as the people of the old East-Asian culture do for their own. We think of those men in whose hands our common fate lies: the Duce, the Tenno [the Japanese emperor] and the *Führer*: a threefold *Sieg Heil*.[11]

Albert Einstein was probably the most outspoken and critical émigré scientist. His letter of 29 April 1933 to Thomas Mann, who was on a lecture tour in Western Europe when he decided not to return to Germany, indicates that only few non-Jewish Germans displayed solidarity or felt compelled to act when Hitler took power: 'I feel urged to tell you what is quite self-evident: the responsible attitude of you and your brother was one of the few rays of hope in the recent events in Germany.'[12] Thomas Mann pointed out that for someone seriously to contemplate breaking away from his country, something unusually wrong and bad probably had to happen, such as this

> German revolution: it lacks all the characteristics that procured the real revolutions, bloody as they may have been, the sympathy of the world. It is by its very nature *not* 'elevation' (*Erhebung*), whatever the proponents may say and shout, but hatred, revenge, primitive desire for manslaughter, and a petty bourgeois soul shabbiness (*Seelenmesquinerie*). Nothing good can come of it, I believe, either for Germany or for the world.[13]

As a consequence of this 'German revolution', German-Jewish scientists were dispersed around the world. The majority of the émigrés in (bio)chemistry (60%) emigrated at first to various countries in Western Europe, particularly the UK (31%). At the end of the 1930s, many émigrés had to emigrate a second time: some had been given only temporary positions in the UK, and many European countries were occupied by the Germans. The United States became the country with the most émigrés in (bio)chemistry (35%), followed by the UK (18%), Switzerland (7%), Palestine (5%), Turkey (4%), Sweden (3%), and Brazil (2%). Of those (bio)chemists who emigrated, around 40 to 50 per cent regained a position in academia – often not equivalent to the one they held before – and another 25 per cent found jobs or posts in industry. At least four academic chemists and an unknown number of industrial chemists were murdered by the Gestapo or deported to concentration camps, where they disappeared (Deichmann 1999).

For the most part, contacts between scientists in Germany and their Jewish colleagues in exile ceased, even in those cases where they had previously been personally friendly. Collaborations stopped, since they were disadvantageous for Germans. Joint publications became impossible, and meetings at international conferences were embarrassing. The biochemist Sir Hans Krebs, a refugee from Freiburg who was awarded a Nobel Prize in 1953, recalled his meetings with scientists from Germany whom he had known before 1933, at the International Congress of Physiological Sciences in Zurich in 1938: 'Some of the German participants seemed embarrassed in the company of refugees. Others were friendly but pretended that nothing had happened, [...] Still others, notably Hans H. Weber, professor of physiology at Münster University and later at Königsberg, with deliberate ostentation associated with the refugees and wanted to be seen in their company' (Krebs 1981: 102). Only a few academics remained faithful to their Jewish teachers and colleagues. One exception was the physiologist Hans Hermann Weber. Here I mention also the renowned chemist Max Vollmer and the Nobel Prize winners Heinrich Wieland and Adolf Windaus, who in many instances refused to go along with the Nazis and helped co-workers or students who, as Jews or political opponents or both, faced severe difficulties, at least as much as was legally possible. The remarkable attitude of the pharmacologist Otto Krayer needs to be emphasized, since he was the only German scientist to refuse the offer of a full professorship formerly held by a Jew (Philipp Ellinger in Düsseldorf), a decision which caused Sir Henry Dale to call him 'the only German gentleman'.[14] Krayer was dismissed immediately from his position as associate professor at the University of Berlin. He first emigrated to England and later became professor at Harvard Medical School in Boston (Goldstein 1987).

Correspondence of German-Jewish (Bio-)chemists
in Exile with their Former Colleagues in Germany

The Situation in Chemistry and Biochemistry at German
Universities after 1945

The general situation in post-war Germany and its universities is well known. Many institutes were damaged or destroyed,
scientific activity, as life in general, suffered from enormous economic
difficulties, and from 1945 to 1947 there was much hunger. At some
institutes work could be continued more or less without interruption,
but at others laboratories and lecture halls had to be rebuilt from
scratch (see e.g. Gerwin 1989; Behrens 1998). This situation should
be emphasized when evaluating the achievements of those scientists
who, under post-war conditions, succeeded in reconstructing their
universities and maintaining teaching and research programmes.

Though West Germany soon benefited strongly from the plan for
economic reconstruction of Europe, launched by the American
secretary of state George C. Marshall as a response to the increase of
military power of the USSR, some scientific fields, in which Germany
had been an international leader in the 1920s, remained backward
even twenty years after the end of the war (for details see Clausen
1964). This was so despite the comparatively fast economic recovery.
The areas of classical organic and inorganic chemistry, which suffered
less than other areas from the expulsion of Jewish scientists and had
achieved much during National Socialism, had no problems making
a rapid recovery. This achievement is reflected in the fact that between
1945 and 1955 four of the chemistry Nobel Prizes (including one for
1944), but none of the physiology and medicine prizes (and only one
of the physics prizes), were given to German scientists. However, a
new start in 1945 was necessary in the biochemistry of intermediate
metabolism, and in the new interdisciplinary fields of molecular
biology, physical-organic chemistry, and polymer chemistry. These
areas had suffered most from the expulsions and also from the rigid
university structures and the traditional dominance of organic
chemistry at German universities. Here the lack of scientific exchange
before, and in particular after, 1945 meant that those fields, some
newly developed in the United States or England, became established
in Germany only after a delay of several years or even decades.

In contrast to the situation after the First World War, there was no
official scientific boycott of German scientists after the Second World
War. But colleagues from former occupied European countries, such

as France and Norway, demanded that German scientists not be invited to international conferences, except for individuals who had remained above reproach. Many foreign scientists as well as German-Jewish émigrés refused to visit Germany (Deichmann 1996: 305–14). The fact that Germany had lost its leading position in most areas of science and, under post-war conditions, could not catch up, only served to enhance the lack of interest among foreign scientists.

It should be noted that after the First World War leading Jewish scientists, both German and non-German, had contributed decisively to the re-establishment of international contacts. These scientists, including Fritz Haber and Otto Meyerhof, were later expelled by the Nazis. An especially dramatic example is that of Ernst Cohen, a physical chemist at the University of Utrecht, who had put great efforts into restoring the relationships of West European scientists with their German colleagues after the First World War (T. Levi 1996). The *Zeitschrift für Physikalische Chemie* honoured him in volume 130 (1927), the Cohen commemorative volume (*Festschrift*) containing the dedication: 'Festband. Herrn Ernst Cohen. Dem erfolgreichen Forscher und unermüdlichen Vorkämpfer für die Wiederherstellung friedlicher Beziehungen zwischen den Gelehrten der durch den Krieg getrennten Völker' (*Festband.* Mr. Ernst Cohen. To the successful researcher and tireless champion of the restoration of peaceful relations between the scholars of peoples divided by the war). Cohen was deported from Holland and murdered in Auschwitz in March 1944 at the age of 74.

After the Second World War there were no longer any Jewish scientists in Germany; many of those who survived in continental Europe had suffered not only from the war but from the anti-Jewish actions in countries under German occupation. Nevertheless, and quite remarkably perhaps, the German biochemist Lothar Jaenicke, persecuted as a half-Jew under the Nazis, claimed that exiled German-Jewish biochemists were of great help to their German colleagues after 1945 (Jaenicke 1989: 72). Who were these scientists who helped their colleagues? Was this attitude widespread among (bio)chemists or confined to individual cases? Did this help indicate a general process of renewal and normalization of scientific and personal relationships among (bio)chemists, despite Auschwitz?

Correspondence

To answer these questions I have examined cases in which contacts were renewed between renowned German-Jewish émigré (bio)chemists and colleagues in Germany after 1945. The early correspondence reveals large individual differences in the way the National Socialist past was perceived, despite common topics and attitudes.[15] In many cases, but not always, the correspondence was initiated from Germany, and the émigrés must often have had in mind thoughts similar to those expressed by Thomas Mann, who received many letters after the Second World War: 'But my joy is a little diminished, not only by the thought that none of them would have been written had Hitler won, but also by a certain lack of knowledge, of empathy, in them, which expressed itself in the naive directness of the renewal, as if these twelve years had never existed.'[16]

The letters written immediately after the end of the Second World War, between 1945 and around 1951, between émigré and non-émigré (bio)chemists can be classified as follows: (1) correspondence with non-Nazi German scientists who were reluctant to face the past; (2) correspondence with German scientists who had publicly endorsed or supported aspects of Nazi policy and now expected help; (3) correspondence that demonstrates help and normalization.[17]

Correspondence with non-Nazi German Scientists who were Reluctant to Face the Past   The prevailing attitude of Germans, not only scientists, after the war was that of being victims – of Hitler, of the war, and of the allied policy of denazification (see below). In correspondence with émigré colleagues, such an attitude can be found particularly among scientists who had not been members of the Nazi party, or who had even despised Nazi ideology.[18] Prominent among them were the chemists and Nobel Prize winners Otto Hahn and Heinrich Wieland and the physicist and Nobel Prize winner Max von Laue. These scientists emphasized the injustice done to German scientists (and Germans in general) by the Allies, were indignant about the range of the denazification programme at German universities, and made an effort to contribute to a better understanding of Germany in the world.

Some émigrés contacted these outspokenly non-Nazi German colleagues in order to regain links with Germany. The correspondence between Otto Hahn and Otto Meyerhof is an example. In a letter to Hahn of June 1946, Meyerhof expressed his 'sympathy and high appreciation' for what Hahn had done in the 'German hell', or had tried to do.[19] Meyerhof, formerly director of the department of physiology of the KWI for Medical Research in Heidelberg, now had a post as visiting professor at the Medical School of the University of Pennsylvania in Philadelphia. He wrote that he had lost many relatives and friends in the gas chambers and concentration camps. Expressing the hope that Hahn's son had returned from the war, he assured him that it was Hahn and von Laue whom foreign colleagues trusted most. Meyerhof had supported the proposal for Hahn's presidency of the Kaiser Wilhelm Society (KWG), a post to which Hahn was appointed in April 1946. Meyerhof wrote that he still took great interest in science in Germany and the KWG, though he could not think of returning to a country that had become a cemetery for those who had been closest to him. Hahn's answer to Meyerhof demonstrates accurately the widespread feelings of Germans who regarded themselves primarily as victims, as well as a lack of sensitivity and his new loyalty as president of the KWG:[20]

> I can understand that after all that had happened to your relatives and friends in Germany you do not feel like (*Sie keine Lust mehr haben*) coming back any more. However, I consider it an injustice on the part of foreign countries to render the majority of the Germans responsible for the events of the past twelve years. If foreign countries knew how incredibly strong the pressure had been, how it was reinforced year by year, then this would enable a certain understanding for the fact that many individuals who were not too strong gave up the fight by saying, 'we cannot change anything'.

The lack of a self-critical note regarding the previous twelve years in this and other letters is striking. Attempts to justify compromises or obedience to Nazi orders (see below) is a common reaction amongst these men; it may reflect certain feelings of guilt or the new and widespread solidarity among Germans, including Nazis, who resented criticism by people whom they thought to be 'other'. In any case, in this and other letters Hahn did not treat Meyerhof as a friend and former German colleague exiled against his wish who now wanted to help him, but as an unwanted critic and an outsider.[21] Referring only to political pressure, Hahn denied the active role of the universities and the KWG in supporting National Socialist anti-Jewish policy. Meyerhof was one of the few émigrés – Carl Neuberg and the physicist Lise Meitner were others – who made efforts to explain to former colleagues in Germany their view of the situation in an effort to assist in bringing about understanding. Meyerhof wrote to Hahn in November 1946:[22]

> I have to say frankly that even the best and most reliable of my German friends, such as you yourself, are not yet liberated from the narrow view,

imposed on them by the Nazis, to such an extent as to be able to analyse the true roots of the situation in Germany and Europe. I fully agree with your complaints about your current situation, but I disagree completely in my interpretation. I stick to the interpretation that Germany owes everything that it is experiencing right now to its *Führer*, and that it owes its *Führer* entirely to itself. Germany has not only been defeated, it has undergone a moral catastrophe without precedent in history. . . Collective judgements are senseless. 'Germany' means to me the majority of responsible and leading persons.

Meyerhof told Hahn that he and other German émigrés stood up for a reasonable policy with regard to German science. But he also made it clear that he demanded of Hahn and other outspoken non-Nazis a much more far-reaching reform of German universities. He did not describe details of this reform but stated that he considered it a prerequisite for any reform to remove from leading positions not only those who had been '150 per cent' Nazis, but also '75 to 99 per cent' Nazis.

In a brief reply Hahn assured Meyerhof 'and many other *American* colleagues' (emphasis added; Hahn showed that he treated Meyerhof as an outsider) of his and von Laue's friendship, but did not comment on basic issues. He had given Meyerhof's letter to von Laue, who wrote a long reply, in which he tried to explain to Meyerhof the German perspective. Thus, von Laue agreed with Meyerhof's judgement of Germany's moral collapse in 1933, and he thought that until mid-1933 opposition would have been possible. But he excused Germany's moral collapse to some extent by stating that other countries were no different, and he even called the dismissal of the large percentage of university teachers by the Allies a continuation of 'Hitler methods'.[23] A few years later, von Laue changed his attitude drastically and became critical not only of the failure of German scientists in 1933 but also of their dishonesty after 1945 (Sime 1996: 364).

It should be noted that Hahn not only represented the KWG/Max-Planck-Gesellschaft (MPG) in post-war Germany but also identified with its decisions in and after 1933. Thus he justified the dismissals of Jewish scientists by the KWG's secretary-generals, Friedrich Glum and Ernst Telschow, by declaring that they had had to obey orders.[24] (In contrast, Glum in his autobiography (1964: 443) spoke of a voluntary subordination to Nazi policies, *Selbstgleichschaltung*, of the KWG in 1933.) When Meyerhof reminded Hahn, among other things, of the loss of his library and scientific apparatus, Hahn suggested that he consider the 'difficulties related to the emigrations', referring

to the difficulties of the KWG. Hahn's loyalty was to Germany and
the KWG/MPG, and not to his dismissed former colleagues. As Ruth
Sime has shown, Lise Meitner was seriously upset by the renewed
nationalist attitude of her old friend and former colleague. On 16
January 1947, she wrote to James Franck about Hahn's visit to Sweden
on the occasion of his receiving the Nobel Prize:

> Hahn said that the Allies are doing in Germany what the Germans did in
> Poland and Russia . . . at which point [Otto] Stern became very upset. He
> absolutely did not respond to our objections; he suppresses the past with
> all his might, even though he always truly hated and despised the Nazis.
> . . . all his subsequent interviews sounded the same. Just forget the past
> and stress the injustice happening now in Germany. As I am part of the
> suppressed past, Hahn never, in any of his interviews about his life work,
> mentioned our long years of work together, nor did he even mention my
> name. (Sime 1996: 345)

It was common in Germany to reject the criticisms of émigrés by
claiming that they did not know or understand the events in Germany,
not only in the immediate post-war period but also much later. Indeed,
Klaus Clusius, a distinguished physical chemist at the University of
Munich, in a letter to Karl Freudenberg in 1946 went so far as to
blame the émigrés for not helping their German colleagues in spite
of knowing what had happened:

> The assaults on professors and universities seem to be of a general nature.
> It seems to be forgotten how science was treated in the Third Reich and
> how universities were castigated (*geschmäht*) because there were hardly any
> professors who had anything to do with Nazism. Should all this be
> forgotten? I can't believe it and at least the German émigrés know enough
> of what happened, even though they did not perceive later developments.[25]

In view of the support given to the Nazi race policy by prominent
anthropologists, the claim that there were hardly any professors who
had anything to do with Nazism is remarkable, as is Clusius's denial
or forgetfulness of the active role of universities in the dismissals.[26]
German émigrés indeed knew exactly what had happened in and after
1933, whereas many non-émigrés obviously preferred to forget it.

The attitudes of Clusius, Hahn, and many others are reminiscent
of the 'sentimentalism, self-pity, selfishness, and the lack of objectivity'
which, as has been described by Michael Balfour and John Mair (1956:
63), 'tend to characterize a nation in defeat'. The huge economic
problems in Germany, a narrow nationalist view, lack of sensitivity,

and, in the case of Hahn, his new role as president of the MPG contributed to this self-centred view. Moreover, at a time when most of their colleagues in other countries had experienced or suddenly become aware of the destruction caused by the war, and of the Holocaust, German scientists, in common with most people in Germany, as Primo Levi wrote (1986: 381), 'wanted *not* to know'.

The question of guilt was thus rarely raised, and if it was, it was immediately generalized and excused. Karl Freudenberg is an example of this tendency. He admitted that every German might be guilty in some way. 'I believe that every decent German has examined which part of the guilt must be referred to him', he wrote to Heinrich Kronstein, the son of an émigré chemist from Karlsruhe, who was then in Washington D.C., on 23 July 1948.[27] He continued:

> I believe that very many have to be very critical of themselves (*sehr ernst mit sich ins Gebet gehen*), and even those who can't be blamed have to say to themselves that they share certain weaknesses and inadequacies (*Untugenden*), which have been inherent in our people. . . . But I want to see the people who could claim to be free of any inadequacies.

Émigrés and Holocaust survivors could not blame the crimes of the Nazi era merely on the weaknesses and inadequacies of the German people. Moreover, many of those whose relatives and friends had been murdered suffered from strong feelings of guilt, because they survived. The latter included the Dutch-American physicist Samuel Goudsmit, whose parents were murdered in Auschwitz (Goudsmit 1947/1996: 48) and the Italian chemist Primo Levi, who survived Auschwitz. In his last book *The Drowned and the Saved*, Levi described

> the feeling of shame or guilt that coincided with reacquired freedom . . . What guilt? When all was over, the awareness emerged that we had not done anything, or not enough, against the system into which we had been absorbed. . . . Consciously or not, [the survivor] feels accused and judged, compelled to justify and defend himself. Self-accusation is more realistic, or the accusation of having failed in terms of human solidarity. (P. Levi 1988: 56–9, 66)

From the correspondence, one notes hardly any feelings of guilt and remorse among German scientists. But many expressed concern about the fate of those colleagues who, as members of the I.G. Farben board of management, were accused or indicted at the Nuremberg trials. They were concerned, as Karl Freudenberg wrote, about 'our friends

Hoerlein and his fellow sufferers' not because 'they had done something unjust but because of the evident intention to dismiss the
justification of their actions', in other words, about the likelihood of
a fair trial.[28] Freudenberg contributed greatly to the defence of I.G.
Farben's chairman Carl Krauch, and Adolf Butenandt to the acquittal
of Heinrich Hoerlein, the director of the Elberfeld branch of I.G.
Farben, who was accused of responsibility for pharmacological
experiments on prisoners in concentration camps.[29]

Here I mention some scientists who either changed their views
through their correspondence or displayed a self-critical attitude
based on their own experience. Among them were the biologist Erwin
Bünning, the aforementioned physiologist Hans Hermann Weber, and
the chemist and Nobel Prize winner Adolf Windaus, all of them non-
Nazis. When Meyerhof reminded Bünning of the moral failure of
German universities and of leading scientists such as Werner Heisenberg and Carl-Friedrich von Weizsäcker, Bünning replied that he had
begun to rethink many things:[30]

> Our isolation from the international community puts us in the seductive
> position of paying more attention to the injustice we are suffering from
> than to the one which took place in our name. It cannot be denied that
> the universities have failed. It is even more regrettable that the under-
> standing of the necessity of denazification is becoming smaller even with
> opponents of the Nazis. . . . I have not kept for myself the thoughts from
> your letters, and I am sincerely thankful to you for arriving at the insight
> that any attempts of justification are totally superfluous, and that only
> positive achievements are able to improve our reputation a little.

Windaus, one of the few scientists who had refused to compromise
with the Nazis (for example, he never used the Hitler greeting), saw
clearly that 'Germany lay in ruins through its own fault', a statement
which was far from obvious to others at the time.[31] In a letter to Carl
Neuberg, he not only described the destruction he saw during a visit
in Berlin, but asked of himself 'what everybody could have done to
prevent the disaster, and what he had failed to do'.

Weber admitted in his letters to Meyerhof, who had been his teacher
at the University of Kiel, that he knew the truth, because during the
war plenty of his former students who had been soldiers visited him
in Königsberg, where he was professor, and told him what they had
seen in Poland. One of them, an army physician, had told him of 'the
annihilation of a complete people by administrative means [such as]
the whole of world history has never witnessed'.[32] Weber concluded,

I used to believe that the German people possesses a particular desire for justice. I have had to realize that this is not true to a particular extent. Moreover, what appeals to the feeling of the Germans appears to them just – and, in addition, true! The sense of morality and reality appears to me to have developed equally badly. Thus Germans have very much abandoned themselves (*preisgegeben*) to lies and moral helplessness and, as a result, self-delusion.

This view brings to mind Hannah Arendt's analysis made during her first post-war visit to Germany. She found a 'widespread flight from reality', expressing itself particularly in the 'inability' and 'unwilling-ness' (*Widerwillen*) to 'distinguish between fact and opinion' (Arendt 1993: 31). According to her and many other émigrés, confronting students with the facts about what had happened was urgently necessary for academic life in Germany. Such a confrontation did not take place, and only decades later did German students start to inquire and ask questions.

Correspondence with German Scientists who had publicly Endorsed or Supported Aspects of Nazi Policy and now Expected Help   After the war, the Allies ordered Germans to undergo a 'denazification' procedure, in which Germans had to declare their previous affiliations with the Nazi party and other Nazi organizations. Civil servants, a category which includes university teachers, were thoroughly investigated and could be dismissed if they had played an active political role after 1933. In their defence they were allowed, however, to present vouchers from well-known people which stated their political innocence. Therefore, many German scientists asked émigrés for help with their 'dena-zification' procedure. These vouchers were called *Persilscheine* (after the name of a German washing powder). Some renowned German scientists did not need *Persilscheine* in order to remain in their positions, but they approached émigrés because they hoped to improve their reputations in the international scientific community. Adolf Buten-andt and Richard Kuhn are prominent examples.

Adolf Butenandt, since 1933 full professor at the Technical University of Danzig, in 1936 was appointed director of the KWI for Biochemistry as successor to Carl Neuberg, who was dismissed in 1934. Also in 1936, he became a member of the Nazi party.[33] During the National Socialist era, he enjoyed major financial support from the DFG and I.G. Farben, in addition to considerable support from the KWG. Like most other researchers, he enjoyed freedom of research, as he pointed out in 1947 to officials of the Rockefeller Foundation: 'Butenandt spoke at some length about the freedom he had been given as a scientist, and the freedom given to the KW Institutes, from the First World War through the Nazi time. Only recently, since the occupation [by the Allies], has this freedom been interfered with.'[34]

During National Socialism Butenandt's research continued to meet high standards. For his isolation and structural elucidation of sex hormones he shared the 1939 Nobel Prize for chemistry with Leopold Ruzicka. He rejected it because the Nazi government did not allow Germans to accept the Nobel Prize after 1935. After the war, Butenandt retained his position as director of the KWI (which in 1944 was transferred to Tübingen), and in addition he became professor of physiological chemistry at Tübingen University. In 1960 he was appointed president of the MPG as successor to Otto Hahn. Carl Neuberg, by contrast, never received a scientific position again after his dismissal in 1934. He emigrated to the United States via various countries, including Palestine, where he arrived in 1938, but where he considered himself too old to live (by then he was 61 years old). From 1936 Neuberg's feelings towards his successor Butenandt and the KWG became increasingly bitter, and after his emigration to New York in 1940, he disseminated among biochemists information about Butenandt's membership of the NSDAP.[35]

Butenandt's correspondence with Neuberg, which started in 1947 and lasted until Neuberg's death in 1956, reveals the different circumstances the two men faced.[36] The possibility cannot be excluded that Butenandt expected advantages from contacting Neuberg. After 1945, Butenandt's refusal to accept a position at Harvard University in 1935, his replacement of a Jewish professor as director of a KWI, and his former membership in the Nazi party rendered his international contacts difficult. For some years, he was not invited to international conferences and did not receive invitations from scientists in the United States, where he had formerly had good scientific connections. In 1951 when Butenandt, who spent some time in the United States, suggested to the *Union Internationale* that he serve as representative for Germany at a meeting in Washington D.C., he did not receive an invitation, even though the Union had just decided to accept Germany again as a member.[37] Without being asked directly, Neuberg offered his assistance. Through his contacts he enabled Butenandt to give guest lectures at American universities. He provided a positive recommendation for Butenandt to the University of Basle, which in 1948 considered Butenandt for a professorship in physiological chemistry. Basle University actually offered Butenandt the professorship despite a negative vote by the Basle parliament. Butenandt declined the offer in 1949.

Butenandt, in return, helped Neuberg receive his pensions from the KWG/MPG, and restored (symbolically) Neuberg's past position at the MPG by placing his photograph among the directors of the KWI for Biochemistry. At Butenandt's suggestion the highest medal of the Federal Republic (*Grosses Verdienstkreuz*) was bestowed on Neuberg in 1954. Neuberg by then was defending Butenandt's behaviour during National Socialism in the United States, where he was the subject of ongoing criticism. In 1954, he went so far as to exonerate Butenandt and the KWG from having been responsible for his (Neuberg's) suffering. On 23 April 1954, he wrote: 'When a free position is offered to somebody, he is allowed to accept it without further thoughts.' Thus shortly before his death, Neuberg bestowed absolution on Butenandt and the MPG.

We can only speculate about why Neuberg acted that way. Despite his experiences in Nazi Germany, Neuberg's emotional attachment to Germany remained strong. He resented what he called an American mentality and lack of culture, and he felt isolated in the United States.[38] Moreover, he admired Butenandt for being the great scientist that he was. In a memorial article about Neuberg, the MPG called Neuberg a victim of a denunciation, portrayed the KWG as a helpless victim of the National Socialists and referred to Neuberg's correspondence with Butenandt as a sign of friendship.[39] Was the latter not rather an expression of the isolation of an expelled Jewish professor?

Like Butenandt, Richard Kuhn was, as mentioned earlier, an outstanding chemist and Nobel Laureate of 1938 (a distinction that he also had to reject). Not having joined the Nazi party, he had no problems remaining in his position as director of the KWI for Medical Research in 1945. However, because he agreed to represent German chemists during National Socialism and also because of his close collaboration with high-ranking Nazi officials, his international reputation was damaged. Until Meyerhof's dismissal as department head of this KWI in 1938, Kuhn and Meyerhof cooperated closely and had a friendly relationship. On 20 July 1945, Kuhn renewed the contact with Meyerhof by a letter in which he asked him whether he would be willing in principle to return to Germany. Meyerhof never learned that Kuhn had denounced his Jewish co-workers in 1936, but he knew that Kuhn had played a leadership role among German chemists under National Socialism. Meyerhof's facilities in Philadelphia were poor compared to those in his institute at the KWI, which, being in Heidelberg, was not destroyed. In a long draft letter dated 11 November 1945, Meyerhof made an attempt to explain to Kuhn his assessment of the situation at a personal level. After calling to mind their different fates, he continued:[40]

> This alone does not separate us, and I do not reproach anybody for making compromises in order to maintain a post and a working place. But you went far beyond that. I cannot withhold the criticism of you by colleagues in the Allied countries that you, by your own free will, put your admirable scientific achievement and your chemical mastery at the service of a regime, of whose inexpressible atrocities and meanness you were well aware. I was particularly hurt, because I knew the liberal atmosphere in which you grew up.

Meyerhof did not post the letter. However, ten months later, on 6 September 1946, he sent Kuhn a response to a letter that Kuhn had sent in mid-August 1946.[41] After expressing thanks for the efforts regarding his restitution, Meyerhof reminded Kuhn that his library, which included original editions of books by Bacon, Newton, Kant, and Fries, had been lost. Meyerhof wrote that he did not understand why Kuhn had not saved the library, for example by convincing the KWG to purchase it. With regard to Kuhn's plans to visit Philadelphia, Meyerhof suggested that he postpone his visit, because due to Kuhn's former political attitude many chemists in the United States had reacted negatively.[42] In contrast to the draft letter, this letter indicates a personal distance from Kuhn.

Meyerhof realized that there was calculation behind Kuhn's proposition that he return. He wrote to his son Gottfried, who had spent some time in Germany as a major in the British army, that Kuhn's proposition had many diplomatic sides, for example the idea that the United States would support the institute once Meyerhof returned. And Meyerhof concluded, 'For many reasons which I don't need to enumerate, I don't think in my dreams of returning to Germany . . . The bonds are cut (*Das Tischtuch ist zerschnitten.*).'[43] In a letter to the American military government in Heidelberg, Meyerhof again emphasized Kuhn's politically liberal background and his great scientific prominence. He suggested that Kuhn's research work should remain unhampered but that he should not be allowed to educate students or represent German chemistry. He closed by rejecting a by then widespread myth among German scientists: 'Probably he still

justifies his former activities by the excuse that he has in this way
saved some scientific values and prevented some more crimes from
being committed. But I don't share this view, which is now accepted
by many German scholars.'[44] Referring to Heisenberg's justification
for his decision not to emigrate but to stay on in Germany despite
the compromises with the Nazis which this implied, Meyerhof's
former student, the émigré biochemist David Nachmansohn (1988:
129, 131), many years later revived this myth.

Correspondence which Demonstrates Help and Normalization   I now return
to Lothar Jaenicke's statement that German-Jewish émigré scientists
helped their colleagues in Germany in the 1950s, that is, at a time
when such help was far from normal. Conference files and correspondence show that a few émigrés did indeed help German colleagues
in various ways. They also reveal tremendous individual differences.

A few émigrés sent out invitations to German colleagues or
postdoctoral fellows and came to Germany to give lectures. However,
such activities did not take place in most cases before the mid-1950s,
but they were indeed of great importance for science in Germany. To
mention a few examples: the biochemist Fritz Lipmann, Nobel
Laureate of 1953, had postdoctoral fellows from many countries in
his laboratory at Rockefeller University in New York, including
Germans, beginning in the late 1950s, most notably Hans Zachau,
later a distinguished molecular biologist. Max Perutz, who left Vienna
in 1937 and emigrated to England as a doctoral student, later
established the Laboratory for Molecular Biology in Cambridge. He
strongly supported Gerhard Braunitzer at the MPI for Biochemistry
in Munich, when Braunitzer worked on the primary structure of
human haemoglobin in the late 1950s.

In April 1933, Ernst Boris Chain, a Jewish chemist from Berlin,
decided to quit his position as research fellow with Peter Rona[45] at
the University of Berlin and to emigrate to England. In cooperation
with Howard W. Florey at Oxford University he successfully isolated
and elucidated the structure of penicillin in the early 1940s, one of
the most important biomedical achievements of this century, which
earned them (together with Sir Alexander Fleming, who had discovered penicillin in 1928) the 1945 Nobel Prize. Chain indicated his
feelings about Germany directly after the war in a letter to Hans
Krebs in November 1945, in which he thanked Krebs for his congrat-
ulations for the Nobel Prize: 'I am profoundly grateful to fate that it
has selected me, a Jew and a refugee from the Nazis, to play a part in

this discovery which has been a benefit to the nations fighting these murderers.'[46]

Robert Ammon, who as assistant to Peter Rona had supervised Chain's doctoral thesis and in 1951 had become professor at the newly founded University at Homburg (Saar), in March 1953 invited Chain to present a keynote lecture at a meeting of the German Physiological-Chemical Society in April 1953.[47] Chain declined, pleading other engagements. A year later, however, Chain came to Germany in order to receive the Paul-Ehrlich/Ludwig-Darmstädter Prize in Frankfurt. In 1957 he accepted an invitation by Ammon to give a lecture about the mode of action of insulin at Homburg University, and his relationship with Ammon became friendly again. In the introduction of his lecture, Chain indirectly referred to the National Socialist past, and he tried to build a bridge to the future through science:[48]

> During the thirty years which have passed since I studied in Berlin, there have been various things in the history of the world, in most cases of a rather unpleasant nature. The life paths of Professor Ammon and myself have taken very different directions, directions none of us would have thought of at the time. However, there is one thing which has remained the same during all these years, and that is our common interest to find out about the tricks of nature. I believe that both of us are amusing ourselves in our laboratories with this project today in the same way as we did twenty-five years ago.

From the 1960s onwards, Chain visited Germany regularly, visiting academic colleagues, among them Otto Westphal, with whom he became friendly, as well as cooperating with industry, in particular Hoechst in Frankfurt, one of the big former I.G. Farben chemical companies. In his letters, Chain rarely mentioned the National Socialist past, of which his mother and sister became victims. Several letters show that he maintained a special relationship with Germany and favourably recalled the cultural and scientific life of Berlin in the 1920s, which was obviously one of his reasons for renewing his cooperation and even friendships with colleagues and companies in Germany. For example, when he was offered the honorary membership of the German Pharmaceutical Society in 1969, he answered:[49] 'Despite all the tragic events of the past, I still have naturally a very special relationship with Germany, in particular with my home city Berlin, and I am concerned to intensify these relationships. I regard the honorary membership of the Pharmaceutical Society as a step in this direction.'

Herman Mark, polymer chemist and founder of the Institute of Polymer Research at Brooklyn Polytechnic Institute, from the 1950s invited Germans to his institute irrespective of their political past, a singular act among German or Austrian émigrés. I learned in the course of several interviews with Mark's former students and colleagues that he had a general tendency to forget about unpleasant things, private or public (for details see Deichmann 1999). This attitude also applied to his contacts with German scholars, some of them known to have been active Nazis. Former colleagues suggested that Mark's extraordinary generosity was at least in part based on the fact that he had not lost any close relatives during the Holocaust. Thus, among the many foreign scholars whom Mark invited to Brooklyn was Kurt Hess, who was a former member of the NSDAP, SA, and SS, and who denounced Lise Meitner to the authorities in order to have her expelled from the KWI for Chemistry in Berlin (Sime 1996: 184–5). Despite the fact that over a long period of time Hermann Staudinger had attacked Mark as a (supposedly) scientific opponent in a way that exceeded the usual tone of scientific controversies (Priesner 1980), Mark invited him to give a lecture in 1957. Before 1938 Mark had a very friendly relationship with Karl Freudenberg, which he renewed after 1945. Directly after the end of war he helped Freudenberg who, due to a denunciation, was jailed in January 1945 (but released shortly after).[50] Mark had Freudenberg elected a member of the advisory board of the *Journal of Polymer Science,* and through his mediation Freudenberg was invited to give guest lectures in the United States, including one before the American Chemical Society in 1951.[51]

Mark's former student and collaborator Herbert Morawetz indicates that one of the driving forces behind Mark's unusual desire to establish and restore good relations with colleagues irrespective of their political past and previous enmities was his wish to be 'the father of polymer chemistry. Therefore he couldn't afford to let any kind of politics interfere with his work. Thus, he also cooperated with the Russians during the period of the Cold War.'[52] Morawetz wrote in a biographical article on Mark (1995):[53]

> Mark's friendship with polymer scientists in all countries, where polymer research was active, made the polymer community an exceptionally close-knit group. Because of his inability to bear grudges, he was most helpful to German and Austrian colleagues at a time when they were frequently ostracized. He also did a great deal to bridge the gulf between the area dominated by the Soviet Union and the rest of the scientific world.

From Sheffield, Hans Krebs, later Sir Hans Krebs, contributed perhaps the most to the advancement of German biochemistry after 1945. Born in Hildesheim in 1900, his parents were assimilated Jews from families which for centuries had lived in Germany. Despite having been deeply hurt by his experiences during and after 1933, Krebs remained, like Meyerhof, deeply attached to German culture, an attitude which helps explain his concern and help for German science. Krebs worked from 1926 to 1930 at the KWI for Cell Physiology in Berlin with Otto Warburg, who was one of the internationally leading biochemists at the time.[54] From 1931 to 1933, Krebs was assistant to Professor Siegfried Thannhauser at the university hospital in Freiburg, a position from which he was dismissed by rector Martin Heidegger, as was Thannhauser a little later. Since Krebs was already famous, due to his discovery of the urea cycle in 1932, he immediately received a position in England and later became professor at the University of Sheffield. In 1953 he shared a Nobel Prize with Fritz Lipmann (see Krebs's biography by Holmes 1991/1993).

Krebs strongly supported – though unsuccessfully at the time – the admission of German participants at the International Biochemistry Congress in Cambridge in 1949.[55] He was one of the earliest post-war visitors to Germany. In 1951 he paid a private visit to Freiburg, and in 1955 he accepted the invitation to deliver the Aschoff Lecture at Freiburg University in memory of the pathologist Ludwig Aschoff. However, it was not an easy decision. Krebs, who knew that his former teacher Thannhauser had refused to return to Freiburg, wrote him on 25 February 1955: 'Perhaps I ought to explain to you the very special circumstances which have prompted me to accept the invitation to give the Aschoff Memorial Lecture.' He rationalized his acceptance by the fact that Aschoff's daughter Eva (like other members of the family) had publicly shown her solidarity with Jews by walking out with him during the May Day parade in 1933, from which Jews were excluded.

Krebs's greatest help to German science, as I see it, lay in his support of two German biochemists, Theodor Bücher and Feodor Lynen. Theodor Bücher, who had not been a member of the Nazi party, and was fourteen years younger than Krebs, had worked as a graduate student with Otto Warburg from 1938 to 1945. In 1946 Krebs helped Bücher publish articles in international journals, supplied him for some years with large numbers of reprints, commented on drafts of Bücher's papers, and made his scientific results known to other colleagues. They also spent many hours together engaged in private discussions. Krebs's help was such that it earned him the comment: 'Your care is that of a big brother', which Bücher meant seriously.[56] Bücher played an important role in re-establishing biochemistry, in particular enzymology, also in its medical application, at German universities (see for example Jaenicke 1989).

Feodor Lynen was the one outstanding German biochemist who after 1945 represented the nearly deserted field of intermediate metabolism, that is, the dynamic part of biochemistry as distinguished from the chemistry of natural products, which flourished in Germany. He was known as an anti-Nazi and excellent scientist, and early on he had friends among colleagues in the United States and England, and also among émigrés. From 1947, Krebs and Lynen began a close collaboration, which some years later also led to a personal friendship. They exchanged their yeast strains, discussed scientific questions extensively, first by letters, and some years later by regular visits. They invited each other to lectures and exchanged students (the first student from England came in 1961). In 1953, Krebs gave an excellent reference for Lynen to Harvard University, which offered him a professorship in biochemistry. Lynen was also offered a professor's post in Germany in that year, and was at first inclined to accept the Harvard offer. However, he decided to remain in Germany. In 1964 Lynen was awarded the Nobel Prize. He was the same age as Krebs when he received the same prize.[57] German biochemistry seemed to have returned to normal.

Conclusion

As these examples indicate, a few émigré biochemists and chemists were of great help to some of their German colleagues in the difficult post-war period. In a few cases a form of normalization (in the above stated meaning) took place, most notably in the relationship between Hans Krebs and Feodor Lynen. However, this did not hold true for German (bio)chemists as a group. A review of the above correspondence makes it clear that a few individual scientists in Germany tried to understand what really happened after 1933, but that there was no collective attempt at understanding and no public expressions of regret. The majority of émigrés and scientists in former German occupied countries expected from their German colleagues what was summed up by Sir Francis (formerly Franz) Simon in 1951. Simon, one of many who declined an invitation by Karl Friedrich Bonhoeffer to participate in a meeting of the Bunsen Gesellschaft in Germany, explained his reasons as follows:[58]

> In my opinion German scientists as a group lost their honour in 1933 and did nothing to get it back. I admit that you cannot say that everybody should have risked his position or life, but such risks were no longer necessary after the war. The least you could expect after all that happened was that German scientists, as a group, would state publicly and clearly that they regretted what had happened. I did not notice anything of this kind.

There was virtually no opposition and no public statement from the KWG or I.G. Farben in 1933 when Jews were dismissed, and there was no such statement in 1945 or later. Nor were émigrés asked to come back except in very few cases.[59] This was one of the reasons why hardly any did come back. A statement of guilt, a hint of regret, at least an acknowledgement that the dismissals were an injustice, was not made at the time. Restitution was granted (if at all) only after a long and humiliating legal procedure as, for example, with the émigré chemists Fritz Arndt and Alexander Schönberg (Singer 1987).60 Now, fifty years later, most of the victims and those responsible are no longer alive. Sadly, German scientists lost the chance for a new beginning in their relations with their former Jewish colleagues in exile. They did not, as a group, take responsibility for the past, a prerequisite for normalization.

Acknowledgements

I thank Ruth Sime, Benno Müller-Hill, Alvin Rosenfeld and Anthony Travis for stimulating comments and their critical reading of the manuscript. The work was supported by a *Habilitations- stipendium* of the Deutsche Forschungsgemeinschaft.

Notes

1. I have reviewed the biographies of 535 academic chemists and bio- chemists who had positions in Germany in 1932, and in Austria and the

German University of Prague in 1937. Their names were extracted from registers of universities and from the annual reports of the Kaiser Wilhelm Society in *Die Naturwissenschaften*. I listed those whose names were absent after these dates and removed those who were absent because of death. Several sources have been used to establish the list of those who were dismissed and those who emigrated. Using the *International Biographical Dictionary of Central European Émigrés, 1933–1945* (ed. H. Strauß and W. Röder, 1983), and other secondary literature, I identified the most prominent émigrés. Information concerning other émigrés was obtained from documents in the Archive of the Max Planck Society (MPG); the list of Displaced German Scholars (of the Notgemeinschaft Deutscher Wissenschaftler im Ausland), which, however, is sometimes erroneous; files of supposed émigrés at the Society of the Protection of Science and Learning in the Bodleian Library in Oxford; the Rockefeller Archive Center in Tarrytown; the collection of the Emergency Committee in Aid of Displaced Foreign Scholars at the New York Public Library; and university archives in Austria and Germany. For methodological details see Deichmann (1999).

2. Apart from smaller files of correspondence, I looked in particular at the extensive correspondence in the papers of Ernst Boris Chain (archive of the Wellcome Institute for the History of Medicine, London), Hans Krebs (archive University of Sheffield), Otto Meyerhof (archive University of Pennsylvania, Philadelphia), Carl Neuberg (American Philosophical Society, Philadelphia), Karl Freudenberg (archive University of Heidelberg), Otto Hahn (archive of the MPG, Berlin), and Richard Kuhn (archive of the MPG, Berlin).

3. I use the term 'normalization' to indicate a process as a result of which the National Socialist past would become irrelevant in regard to scientific communication and cooperation. This usage would imply, for example, the renewal of previous contacts or the establishment of new exchanges and collaboration between German-Jewish émigré scientists and colleagues in Germany without reservations concerning the political past, mutual invitations to guest lectures and an exchange of (post)doctoral students.

4. Author's interview with Erwin Chargaff, New York, 28 January 1997.

5. For the usage of the Science Citation Index and its limitations, see Deichmann (1999).

6. Karl Freudenberg to George Barger, 17 July 1933, Freudenberg papers, rep. 14/111, translation UD.

7. See Freudenberg papers, Rep. 14.

8. Paul Harteck papers, Rensselaer Polytechnic Institute, Troy, part 1 MC 17, box 1, translation UD.

9. The political, economic and military role which Nazi science politicians attributed to basic science has been analysed in Deichmann (1996: 105–31).

10. Richard Kuhn to Friedrich Glum, 27 April 1936, archive of the MPG, Abt. 1, Rep. IA, 540/2.
11. Translation UD. A detailed analysis of German chemists' collaboration with the Nazi state and party and their research is given in *Chemie – Innenansicht einer Wissenschaft 1933–1945* by the author, forthcoming. For Kuhn's war-related research see: 'Kriegsbezogene biologische, biochemische und chemische Forschung an den Kaiser Wilhelm-Instituten für Züchtungsforschung; Physikalische Chemie und Elektrochemie; Medizinische Forschung', by the author, forthcoming. In some instances Kuhn's attitude shows similarities to that of the physicist and Nobel Prize winner Werner Heisenberg, one of the most eminent German physicists of the twentieth century. Like Kuhn not a card-carrying member of the NSDAP, Heisenberg agreed to be a representative of Germany in occupied Denmark and the Netherlands and, among other things, tried to convince his foreign colleagues that the German reordering of Europe was a necessary evil (Cassidy 1992: 464–73; Walker 1990: 131–46). This Europe, of course, would be a Europe without Jews. Because of this, Michael Frayn, in his recent play *Copenhagen*, misses the point by claiming that what was at stake during Heisenberg's 1941 visit to Bohr in Copenhagen was the question of the responsibility of scientists with regard to the atomic bomb. In fact that was Heisenberg's post-war apologia. At the time most of Heisenberg's colleagues, including Bohr, thought that Heisenberg was probing him for information on the Allied war effort. The biochemist Erwin Chargaff compared Richard Kuhn with Heisenberg: 'Essentially he was very good, but he degenerated, politicized in the same way as Heisenberg, who, in spite of that, came closest to being a genius.' Chargaff called Kuhn a 'Karajan of Chemistry' (interview with author, 28 January 1997). The conductor Herbert von Karajan had good connections with high-ranking Nazi politicians; in addition, he became (unlike Kuhn) a member of the NSDAP.
12. Albert Einstein to Thomas Mann, 29 April 1933, A. Einstein Archives, Hebrew University of Jerusalem, translation UD.
13. Thomas Mann to Albert Einstein, 15 March 1933, ibid., translation UD.
14. Erwin Chargaff in interview with author, 28 January 1997, translation UD.
15. The group of renowned émigré (bio-)chemists considered here (see above) is not representative of all émigrés, and the readiness of some of them to discuss at length problems related to the past is rather exceptional, but the discussions reveal typical attitudes.
16. Thomas Mann to Walter von Molo, Warum ich nicht nach Deutschland zurückkehre, *c.* 1945, in: *Reden und Aufsätze* 4, Gesammelte Werke Bd. XII, (S. Fischer-Verlag, Frankfurt 1974: 957), translation UD.

17. For the usage of 'normalization', see n. 3. The first two groups were by far the largest ones. In the case of Lise Meitner, too, most letters she received between 1945 and 1950 were either from younger, former National Socialist colleagues who asked her for political affidavits or from non-Nazi German friends, such as Hahn or von Laue, who looked at the problems entirely from the German point of view (Sime 1996: 347–61). The only one of her former German colleagues fully to understand her was Fritz Straßmann, who was also the only one to make efforts to find a position for her in Germany.

18. For former National Socialists it was not opportune to contradict émigrés, because they needed their help (see p. 000).

19. Meyerhof to Hahn, 25 June 1946, Meyerhof papers, translation UD.

20. Hahn to Meyerhof, 5 August 1946, ibid., translation UD.

21. For Hahn's attitude in Stockholm at the Nobel ceremonies in November 1946, see above (p. 000). However, it is interesting that Hahn's letters to colleagues in Germany show a somewhat different picture. For example, on 4 May 1946 he asked Klaus Clusius, who then was the dean of the faculty of natural sciences at the University of Munich, to what extent it was true that a large percentage of active Nazis had been reinstated at this university, a concern he did not share with his colleagues in exile (Hahn papers).

22. Meyerhof to Hahn, 8 November 1946, Hahn papers, Rep. III/14A/2937, translation UD.

23. von Laue to Meyerhof, 24 December 1946, Meyerhof papers, translation UD.

24. Hahn to Meyerhof, 4 June 1948, ibid.

25. Klaus Clusius to Karl Freudenberg, 3 March 1946, Freudenberg papers, Rep.14/147, translation UD. Other German scientists after the war blamed émigrés for not having experienced National Socialism themselves. Here I add a personal experience. I used as an afterword to my book *Biologists under Hitler* (1996) a letter by Lise Meitner to Otto Hahn, which she wrote immediately after the photographs of concentration camp victims had become widely known outside Germany and in which she tried to explain to Hahn as a friend what she and other people outside Germany were feeling. I was strongly criticized by some German scientists for putting this letter at the end of my book, and one of them in particular criticized not only me for making judgements about National Socialism, which I had not experienced myself, but also Lise Meitner for the same reason. This scientist had been an active member of National Socialist organizations, and he had, no doubt, obvious reasons for this attitude.

26. Remarkably, the first publication to reveal the massive support of Nazism by German academics (Max Weinreich, *Hitler's Professors*, 1946), has not been translated into German.

27. Freudenberg papers, Rep. 14/212, translation UD.
28. Freudenberg to Paul Karrer, 8 December 1947, ibid., translation UD.
29. In a letter of 25 June 1948 from Nuremberg prison, Krauch thanked Freudenberg for 'his immense and valuable help, by which he had contributed to his defence' (Freudenberg papers). According to Peter Karlson (1990: 149–51), Butenandt's testimony at the Nuremberg war crimes trials contributed decisively to the acquittal of Hoerlein.
30. Erwin Bünning to Otto Meyerhof, 15 February 1948, Meyerhof papers, translation UD.
31. Adolf Windaus to Carl Neuberg, 22 June 1947, Neuberg papers, translation UD.
32. Hans H. Weber to Otto Meyerhof, 29 October 1946, Meyerhof papers, translation UD.
33. 1 May 1936, No. 3716562, Bundesarchiv Berlin, formerly Berlin Document Center, file A. Butenandt.
34. Robert J. Havighurst, 6 October 1947, Rockefeller Archive Center, Tarrytown, Files of the Rockefeller Foundation.
35. Salome Glücksohn-Waelsch in interview with author, New York, 21 October 1991.
36. Butenandt's exact motives in resuming contact with Neuberg cannot be determined, for example whether a feeling of guilt might have played a role. Thus in February 1947, Butenandt suggested to Hahn, the newly elected president of the MPG, that he offer James Franck a position at a German MPI or university and the possibility to work in Hechingen, not only because it would serve the interests of the Hechingen institute but also because it would be a 'wonderful possibility to make good with Franck an injustice which strongly worries all of us' (Hahn papers). Even though Butenandt remarked about injustice only with respect to the particular case of James Franck, and did so in a private letter, he was one of very few German scientists at the time to call the dismissals in 1933 an injustice at all. The following quotations are in the Butenandt–Neuberg correspondence, Neuberg papers.
37. Adolf Butenandt to Karl Freudenberg, 28 November 1951, Freudenberg papers. Rep.14/120.
38. Evidence can be found for example in many of Neuberg's letters, in particular to German colleagues (Neuberg papers).
39. MPG-Spiegel 6/1983.
40. Meyerhof papers, translation UD.
41. Meyerhof to Kuhn, 6 September 1946, Kuhn papers, Archive MPG, Rep. III/25/54.
42. Kuhn planned to visit Philadelphia because Paul György, a former co-worker of his and an émigré medical scientist in Philadelphia, invited him. György tried to get Kuhn to Philadelphia permanently and in 1949 Kuhn was offered a position as professor for physiological chemistry at

the University of Pennsylvania in Philadelphia (which Kuhn declined in 1951). Industry, in particular the company Rohm and Haas, was involved in this invitation. However, the attitude of the university towards Kuhn became increasingly negative. Other émigrés spread information about Kuhn's compromises with the Nazis, and about his speech to the German Chemical Association in 1942 (see p. 000) (note of Hans Krebs about discussions with Otto Westphal at Hamburg on 29 July 1976, Krebs papers).

43. Otto Meyerhof to Gottfried Meyerhof, 4 November 1945, Meyerhof papers, translation UD.
44. Otto Meyerhof to military government in Heidelberg, 29 January 1947, ibid.
45. Peter Rona moved after his dismissal to Budapest, where he was born. Although protected by the Swedish embassy from deportation, he died in 1944, either murdered by Germans or having committed suicide.
46. Chain to Krebs, 23 November 1945, Krebs papers.
47. The following information and quotations are from correspondences in the papers of E. B. Chain.
48. Chain, *Der Wirkungsmechanismus des Insulins*, Chain papers, translation UD.
49. Chain to F. Schenck, 28 July 1969, Chain papers. For Chain's life and work, see Clark 1985.
50. On 26 July 1946 Freudenberg thanked Mark for his letter, which, he said, had been a great help. Freudenberg was acquitted of all accusations (Freudenberg papers, Rep 14/231).
51. Freudenberg papers, Rep 14/147.
52. Herbert Morawetz in interview with the author, New York, 26 September 1996.
53. Further information on Mark's life and work can be found in Bohning and Sturchio (1986) and Mark (1993).
54. To my knowledge Warburg was the only 'half'-Jewish scientist who kept his position as director of a KWI throughout National Socialism; this was possible because his institute and research were almost entirely funded by the Rockefeller Foundation, and he had access to high-ranking Nazis who protected him (see e.g. Albrecht/Hermann 1990: 371; Höxtermann/Sucker 1989: 142; Müller-Hill 1984: 32).
55. This and the following information and quotations are from Krebs's papers and correspondences.
56. Bücher to Krebs, 8 March 1952, ibid.
57. Information on Lynen's life and work can be found in Decker (1984).
58. Franz Simon to Karl-Friedrich Bonhoeffer, 22 March 1951, archive of the MPG, Berlin, Nachlass KF Bonhoeffer, translation UD.
59. It is true that many German universities and cities were severely damaged, and that the economic situation directly after the war was

very bad. Even though many émigrés would not have returned for this
and other reasons, they regarded it as a new form of injustice not to
even have been asked (see Deichmann, *Chemie – Innenansicht*: 540). In
particular, many émigré scientists in Turkey strongly hoped that they
would be offered positions in Germany, and they were ready to return
because of 'cultural hunger', despite knowing about the situation in
Germany (F. Breusch to Otto Meyerhof, 14 January 1946, Meyerhof
papers).
60. For a survey on restitutions in the Federal Republic of Germany, see
Herbst/Goschler (1989) and Goschler (1992).

References

Albrecht, Helmuth and Hermann, Armin (1990), 'Die Kaiser-
Wilhelm-Gesellschaft im Dritten Reich (1933–1945)', in Rudolf
Vierhaus and Bernhard vom Brocke (eds), *Forschung im Spannungsfeld
von Politik und Gesellschaft. Geschichte und Struktur der Kaiser-Wilhelm-/
Max-Planck-Gesellschaft aus Anlaß ihres 75jährigen Bestehens*, Stuttgart:
dva, 356–406.
Arendt, Hannah (1950/1993), *Besuch in Deutschland*. Neuauflage des
1950 veröffentlichten Essays, Berlin: Rotbuch-Verlag.
Balfour, Michael and Mair, John (1956), *Four-Power Control in Germany
and Austria*, London/New York: Oxford University Press.
Behrens, Helmut (1998), *Wissenschaft in turbulenter Zeit. Erinnerungen
eines Chemikers an die Technische Hochschule München 1933–1953*,
München: Münchener Universitätsschriften.
Beyerchen, Alan (1977), *Scientists under Hitler*, New Haven: Yale
University Press.
Bohning, James J. and Sturchio, Jeffrey (1986), *Mark, Herman. Oral
History, Transcript of interviews at Polytechnic University on 3 February,
17 March and 20 June 1986*: The Beckman Center for the History of
Chemistry.
Cassidy, David (1992), *The Life and Science of Werner Heisenberg*, New
York: Freeman and Co.
Clark, Ronald W. (1985), *The Life of Ernst Chain. Penicillin and Beyond*,
New York: St. Martin's Press.
Clausen, Richard (1964), *Stand und Rückstand der Forschung in
Deutschland*, Wiesbaden: Steiner.
Decker, K. (1984), 'Feodor Lynen 1911–1979', *Liebig's Annalen der
Chemie*, 9: i–xl.

Deichmann, Ute (1996), *Biologists under Hitler*, Cambridge MA: Harvard University Press.

—— (1999), 'The Expulsion of Jewish Chemists and Biochemists from Academia in Nazi Germany', *Perspectives on Science*, 7: 1–86.

Fischer, Klaus (1988), 'Der quantitative Beitrag der nach 1933 emigrierten Naturwissenschaftler zur deutschsprachigen physikalischen Forschung', *Berichte zur Wissenschaftsgeschichte*, 11: 83–104.

Fölsing, Albrecht (1993), *Albert Einstein. Eine Biographie*, Frankfurt a. M.: Suhrkamp.

Freudenberg, Karl (1940), '"Buchbesprechungen": Meyer, K. H. und H. Mark: Hochpolymere Chemie. Ein Lehr- und Handbuch für Chemiker und Biologen. Bd. I. H. Mark: Allgemeine Grundlagen der hochpolymeren Chemie. Leipzig 1940', *Die Naturwissenschaften*, 28: 579.

Gerwin, Robert (ed.) (1989), *Wie die Zukunft Wurzeln schlug*, Berlin: Springer.

Glum, Friedrich (1964), *Zwischen Wissenschaft, Wirtschaft und Politik. Erlebtes und Erdachtes in vier Reichen*, Bonn: Bouvier.

Goldstein, Avram (1987), 'Otto Krayer. October 22, 1899–March 18, 1982. Biographical Memoirs', *Nat. Acad. of Sciences*, 57: 151–225.

Goschler, Constantin (1992), *Wiedergutmachung. Westdeutschland und die Verfolgten des Nationalsozialismus, 1945–1954* (Quellen und Darstellungen zur Zeitgeschichte 34), München: Oldenbourg.

Goudsmit, Samuel A. (1947/1996), *Alsos*, 3rd edn., Woodbury: American Institute of Physics.

Herbst, Ludolf and Goschler, Constantin (eds) (1989), *Wiedergutmachung in der Bundesrepublik Deutschland*, München: Oldenbourg.

Höxtermann, Ekkehard and Sucker, Ulrich (1989), *Otto Warburg* (Biographien hervorragender Naturwissenschaftler, Techniker und Mediziner 91), Leipzig: Teubner.

Holmes, Frederick L. (1991/1993), *Hans Krebs, vol. 1: The Formation of a Scientific Life, 1900–1933*; vol. 2: *Architect of Intermediary Metabolism, 1933–1937*, Oxford: Oxford University Press.

Jaenicke, Lothar (1989), 'Wieviel Zufälliges doch in der Entwicklung steckt', in Robert Gerwin (ed.), *Wie die Zukunft Wurzeln schlug. Aus der Forschung der Bundesrepublik Deutschland*, Berlin: Springer, 31–8.

Karlson, Peter (1990), *Adolf Butenandt. Biochemiker, Hormonforscher, Wissenschaftspolitiker*, Stuttgart: Wissenschaftliche Verlagsgesellschaft.

Krebs, Hans (1981), *Reminiscences and Reflections*, Oxford: Oxford University Press.

Levi, Primo (1986), *Survival in Auschwitz and the Reawakening: Two Memoirs*, New York: Summit.

—— (1988), *The Drowned and the Saved*, London: Michael Joseph.

Levi, Tansjö (1996), 'Die Wiederherstellung von freundschaftlichen Beziehungen zwischen Gelehrten nach dem 1.Weltkrieg. Bestrebungen von Svante Arrhenius und Ernst Cohen', in Gerhard Pohl (ed.), *Naturwissenschaften und Politik. Tagungsband zur Vortragstagung an der Universität Innsbruck*, April 1996: 72–80.

Mark, Herman (1993), *From Small Organic Molecules to Large: A Century of Progress*, Washington, D.C.: American Chemical Society.

Morawetz, Herbert (1995), 'Herman Mark. Life and Accomplishments', *Macromol. Symp.*, 98: 1173–84.

Müller-Hill, Benno (1984), *Tödliche Wissenschaft. Die Aussonderung von Juden, Zigeunern und Geisteskranken 1933–1945*, Reinbek: Rowohlt.

Nachmansohn, David (1979), *German-Jewish Pioneers in Science 1900–1933*, New York: Springer.

—— (1988), *Die große Ära der Wissenschaft in Deutschland 1900 bis 1933. Jüdische und nichtjüdische Pioniere in der Atomphysik, Chemie und Biochemie*, aus dem Englischen überarbeitet und erweitert von Professor Dr. Roswitha Schmid, Stuttgart: Wissenschaftliche Verlagsgesellschaft.

Nordwig, Arnold (1983), 'Vor fünfzig Jahren: der Fall Neuberg. Aus der Geschichte des Kaiser-Wilhelm-Instituts für Biochemie zur Zeit des Nationalsozialismus', *MPG-Spiegel*, 6: 49–53.

Priesner, Claus (1980), *H. Staudinger, H. Mark, K.H. Meyer. Thesen zur Grösse und Struktur der Makromoleküle*, Weinheim: Verlag Chemie.

Sime, Ruth (1996), *Lise Meitner. A Life in Physics*, Berkeley: University of California Press.

Singer, Erich (1987), 'Alexander Schönberg 1892–1985', *Chem. Ber.*, 120, issue 9: i–xix.

Strauss, Herbert A. and Röder, Werner (eds) (1983), *International Biographical Dictionary of Central European Emigrés 1933–1945*, vol. II in two parts, München: K. G Saur.

Szöllösi-Janze, Margit (1998), *Fritz Haber 1868–1934. Eine Biographie*, München: C. H. Beck.

Walker, Mark (1989), *German National Socialism and the Quest for Nuclear Power 1939–1949*, Cambridge: Cambridge University Press.

—— (1990), *Die Uranmaschine. Mythos und Wirklichkeit der deutschen Atombombe*, Berlin: Siedler.

Weinreich, Max (1946), *Hitlers Professors. The Part of Scholarship in Germany's Crimes against the Jewish People*, New York.

Wise, Norton (1994), 'Pascual Jordan: Quantum Mechanics, Psychology, National Socialism', in Monika Renneberg and Mark Walker (eds), *Science, Technology, and National Socialism*, Cambridge: Cambridge University Press, 224–54.

Notes on Contributors

UTE DEICHMANN is historian of science at the Institute of Genetics at the University of Cologne, where she is Associate Professor (Privatdozentin). Her research projects have been supported by the Deutsche Forschungsgemeinschaft. She has published extensively on the history of biology, chemistry and biochemistry under National Socialism. Her books include *Biologen unter Hitler* (1992/1995; translated as *Biologists under Hitler* 1996) and her Habilitationsschrift *Chemie – Innenansicht einer Wissenschaft 1933–1945* (forthcoming). She was awarded the Ladislaus Laszt International and Social Concern Award of Ben Gurion University of the Negev in Beer Sheva, Israel, in 1995. In 1996/7, she was appointed Edelstein International Fellow in the History of Chemical Sciences and Technologies.

STEFAN KÜHL has degrees in both history and sociology. After his studies at the universities of Bielefeld, Johns Hopkins, Baltimore, Paris and Trinity College, Oxford, he focused on organizational sociology at the universities of Bangui (Central African Republic) and Magdeburg. He is presently Research Fellow at the Institute of Sociology, University of Munich. His numerous publications include *The Nazi Connection. Eugenics, American Racism and German National Socialism* (1994) and *Die Internationale der Rassisten. Aufstieg und Niedergang der internationalen Bewegung für Eugenik und Rassenhygiene im 20. Jahrhundert* (1997).

LUITGARD MARSCHALL earned a degree in pharmaceutics from the University of Freiburg and has since researched into the history of science and technology, especially the history of biotechnology in Germany. Her study *Im Schatten der Synthesechemie. Industrielle Biotechnologie in Deutschland 1900–1970* (2000) received the Rudolph-Kellermann-Preis für Technikgeschichte 1997 and the ABB-Wissenschaftspreis 1998. She has been awarded several scholarships, amongst others by the Deutsches Museum in Munich and the Institute for European History in Mainz. She is presently Research Fellow for the History of Technology at the Munich Center for the History of Science and Technology.

SYLVIA PALETSCHEK is Associate Professor (Hochschuldozentin) for Modern History at the University of Tübingen. Her main fields of research are gender history, the revolutions of 1848/9, and the social history of modern universities in Germany. Her numerous publications include *Frauen und Dissens. Frauen im Deutschkatholizismus und in den freien Gemeinden 1841–1852*

(1990), *Women's Emancipation Movements in Europe in the Long Nineteenth Century* (co-editor, 2001) and *Die permanente Erfindung einer Tradition. Studien zur Geschichte der Universität Tübingen im Kaiserreich und in der Weimarer Republik* (2000).

CAY-RÜDIGER PRÜLL is Associate Professor (Privatdozent) for History of Medicine and is working as Senior Research Associate at the University of Durham. He has degrees in both history and medicine. His main research topic is the social and cultural history of modern medicine, in particular the histories of pathology and psychiatry. He is participating in a research project on the impact of the receptor concept on 20th century pharmacology. His numerous publications include *Der Heilkundige in seiner geographischen und sozialen Umwelt – Die Gießener Medizinische Fakultät auf dem Weg in die Neuzeit* (1993), *Pathology in the 19th and 20th Centuries* (editor, 1998) and *Medizin am Toten oder am Lebenden? Pathologie in Berlin und in London 1900 bis 1945* (Habilitation-thesis, forthcoming).

JOSEF REINDL read Modern History and Economic History at the University of Munich, the London School of Economics and at St Antony's College, Oxford. His fields of research cover a broad range from the history of telegraphy to comparative business history and the contemporary history of medical sciences. Among his main publications are *Der Deutsch–Österreichische Telegraphenverein und die Entwicklung des deutschen Telegraphenwesens 1850–1871* (1993) and *Wachstum und Wettbewerb in den Wirtschaftswunderjahren. Die britische und deutsche elektrotechnische Industrie* (2000). At present Research Fellow at the University of Munich, he is preparing a study on *Biomedizinische Forschung in Berlin-Buch, 1928–1998*.

MECHTILD RÖSSLER earned degrees in geography from the universities of Freiburg and Hamburg. After several Research Fellowships at the universities of Hamburg, Paris and Berkeley, she has been working at UNESCO in Paris since 1991. She has published widely, in particular on the history of geography during the Third Reich. Her research work has been supported by the Deutsche Forschungsgemeinschaft. Among her numerous publications are *'Wissenschaft und Lebensraum'. Geographische Ostforschung im Nationalsozialismus. Ein Beitrag zur Disziplingeschichte der Geographie* (1990) and *Der Generalplan Ost. Aspekte der nationalsozialistischen Bevölkerungs- und Vernichtungspolitik* (co-editor, 1993).

MARGIT SZÖLLÖSI-JANZE is presently Professor for Contemporary History at the University of Salzburg, Austria. Apart from the history of fascism, her main field of research is the history of science and technology in the context of general history. Her books include *Die Pfeilkreuzlerbewegung in Ungarn. Historischer Kontext, Entwicklung und Herrschaft* (1989) and *Geschichte der Arbeitsgemeinschaft der Großforschungseinrichtungen 1958–1980* (1990). For her biography on the German-Jewish Nobel Prize winner Haber (*Fritz Haber 1868–1934 – Eine Biographie,* 1998) she was awarded the Prize of the University of Munich and the Prize of the German Historians' Association. As the Stifterverband Visiting Fellow at St Antony's College, Oxford, in 1998/9, she researched the history of the sciences under National Socialism and organized the seminar series on which the present volume is based.

HELMUTH TRISCHLER is Director of the Research Department of the Deutsches Museum and Professor for Modern History at the University of Munich. His main fields of research are social history and the history of science and technology in the nineteenth and twentieth centuries. Among his numerous publications are books like *Steiger im deutschen Bergbau. Zur Sozialgeschichte der technischen Angestellten 1815–1945* (1988), *Luft- und Raumfahrtforschung in Deutschland 1900–1970. Politische Geschichte einer Wissenschaft* (1992) and *Forschung für den Markt – Geschichte der Fraunhofer-Gesellschaft* (co-author, 1999).

Index